71 Topics in Current Chemistry
Fortschritte der Chemischen Forschung

Inorganic Chemistry Metal Carbonyl Chemistry

Springer-Verlag
Berlin Heidelberg GmbH 1977

This series presents critical reviews of the present position and future trends in modern chemical research. It is addressed to all research and industrial chemists who wish to keep abreast of advances in their subject.

As a rule, contributions are specially commissioned. The editors and publishers will, however, always be pleased to receive suggestions and supplementary information. Papers are accepted for "Topics in Current Chemistry" in English.

ISBN 978-3-662-15841-8 ISBN 978-3-540-37328-5 (eBook)
DOI 10.1007/978-3-540-37328-5

Library of Congress Cataloging in Publication Data. Main entry under title: Inorganic chemistry. Metal carbonyl chemistry. (Topics in current chemistry ; 71) 1. Metal carbonyls--Addresses, essays, lectures. I. Series.
QD1.F58 no. 71 [QD411] 540'.8s [547'.05] 77-9034

Originally published by Springer-Verlag Berlin Heidelberg New York in 1977
Softcover reprint of the hardcover 1st edition 1977

2152/3140-543210

Contents

Springer-Verlag	Postfach 105 280 · D-6900 Heidelberg 1 Telephone (0 62 21) 4 87-1 · Telex 04-61 723 Heidelberger Platz 3 · D-1000 Berlin 33 Telephone (0 30) 82 20 01 · Telex 01-83319
Springer-Verlag New York Inc.	175, Fifth Avenue · New York, NY 100 10 Telephone 477-82 00

Tetranuclear Carbonyl Clusters

Prof. Paolo Chini

Istituto di Chimica Generale dell'Università, Via G. Venezian 21, 20133 Milano, Italy

Dr. Brian T. Heaton

The Chemistry Laboratory, University of Kent at Canterbury, Canterbury, Kent, England

Table of Contents

I. Introduction

We are pleased to accept the invitation of the editors to present a review on metal carbonyl clusters which affords us the opportunity to discuss the tetranuclear clusters and to complete the previous discussion of 3-dimensional clusters[60)].

In the literature there is much confusion over the terminology used for polynuclear, cluster and cage compounds. Although the present use of the term cluster to describe compounds containing metal-metal bonds is widespread, it is an exaggeration to describe dinuclear compounds as clusters and it is particularly inappropriate to call a cage compound, in which there are no metal-metal bonds, a cluster compound. Considering the arrangement of the metal atoms, a 3-dimensional cluster may be defined as a 3-dimensional net-work of metal atoms which is held together by metal-metal bonds, and within which each metal atom forms at least three different metal-metal bonds. In this type of structure a pronounced similarity with the metallic state is expected.

Actually at nuclearity four, which represents the border between 2- and 3-dimensional structures, both types of structures are observed, and therefore in the present review we will discuss all transition metal tetranuclear carbonyl systems in which there are metal-metal bonds. We hope to have covered the literature comprehensively up to the end of 1975.

In the text and in the tables we will use the usual abbreviations. The less usual are the following:

Bz = benzyl
Cp = (η^5-C_5H_5)
Cy = Cyclohexyl
DEF = diethylformamide
DMF = dimethylformamide
DPE = 1,2-bis(diphenylphosphino) ethane
DPM = 1,2-bis(diphenylphosphino) methane
ETPO = 4-ethyl-2,6,7-trioxa-1-phosphabicyclo[2.2.2.]-octane
mes = mesitylene
m-Tol = *m*-tolyl, (*o* = ortho, *p* = para)
α-PE = 1,2-di-α-pyridylethylene
PPN^+ = bis(triphenylphosphino)iminium cation
ffars = 1,2-bis(dimethylarseno)tetrafluorocyclobutene

The first tetranuclear cluster, $Co_4(CO)_{12}$, was prepared in 1910 by Mond, Hirz and Cowap from the pyrolysis of $Co_2(CO)_8$ [195)], and its structure in the solid state was clarified between 1959 and 1965 (Corradini[71)], and Wei and Dahl[241)]. In 1957 Hieber and Werner[119)] isolated the first tetranuclear carbonyl anions. In 1959 the first mixed carbonyl, $FeCo_3(CO)_{12}H$, was reported[51)] and this opened up the large field of clusters containing different metals. Between 1967 and 1968 the possibility of substituting carbon monoxide with other ligands emerged[48, 174)]. As shown in Tables 1 and 2, more than 90 not trivially different examples of tetranuclear clusters are presently known, the majority of which are 3-dimensional and tetrahedral.

Table 1. Homonuclear tetranuclear carbonyl clusters

	$Fe_4(CO)_{13}H_2$ $[Fe_4(CO)_{13}H]^-$ $[Fe_4(CO)_{13}]^{2-}$ $[Fe_4(CO)_4Cp_4]^n$ (n = 2+, 1+, 0, −1)	$Co_4(CO)_{12}$ $Co_4(CO)_{12-n}L_n$ n = 1,2,3,4,5 $Co_4(CO)_9$(arene) $Co_4(CO)_8$ (arene) L $[Co_4(CO)_{11}(COOR)]^-$ $Co_4(CO)_{10}(C_2R_2)$ $Co_4(CO)_{10}(CS_2)$ $Co_4(CO)_{10}E_2$ E=S, Te, PPh $Co_4(CO)_4(SEt)_8$ $Co_4(CO)_2Cp_4$	$Ni_4(CO)_6L_4$ $Ni_4(CO)_4(C_4F_6)_3$
	$Ru_4(CO)_{13}H_2$ $[Ru_4(CO)_{12}H_3]^-$ $Ru_4(CO)_{12}H_4$ $[Ru_4(CO)_{12}(RC_2R')H]^+$ $Ru_4(CO)_{12}(RC_2R')$ $Ru_4(CO)_{11}(RC_2R')(PhC_2R'')$ $Ru_4(CO)_{12-n}L_nH_4$ n = 1,2,3,4 $Ru_4(CO)_{12}(C_8H_{10})$ $Ru_4(CO)_{12}(C_8H_{12})$ $Ru_4(CO)_{11}(C_8H_{10})$ $Ru_4(CO)_{10}(C_{12}H_{16})$ $Ru_4(CO)_9$(azulene-Me_3) $Ru_4(CO)_4Cp_4$ $Ru_4(CO)_6Cp_3$	$Rh_4(CO)_{12}$ $Rh_4(CO)_{12-n}L_n$ n = 1,2,3,4 $[Rh_4(CO)_{11}(COOR)]^-$ $[Rh_4(CO)_{11}]^{2-}$ $Rh_4(CO)_{10}(C_2R_2)$ $Rh_4(CO)_8(COT)$ $Rh_4(CO)_2Cp_4$	
$[Re_4(CO)_{16}]^{2-}$ $[Re_4(CO)_{15}H_4]^{2-}$ $[Re_4(CO)_{16}(OCH_3)H_4]^{3-}$ $Re_4(CO)_{12}[\mu_3\text{-InRe}(CO)_5]_4$ $[Re_4(CO)_{12}H_6]^{2-}$ $Re_4(CO)_{12}H_4$ $Re_4(CO)_{10}(PPh_2Me)_6$	$Os_4(CO)_{13}H_2$ $Os_4(CO)_{12}H_4$	$Ir_4(CO)_{12}$ $Ir_4(CO)_{12-n}L_n$ n = 1,2,3,4 $[Ir_4(CO)_{12}H_2]^{2+}$ $[Ir_4(CO)_{11}X]^-$ X=H, COOR $[Ir_4(CO)_{10}H_2]^{2-}$	$Pt_4(CO)_5L_4$ $Pt_4(CO)_5L_3$ $Pt_4(CO)_{8-6}L_3$

Table 2. Mixed Tetranuclear Carbonyl Clusters (each compound is discussed in the section shown in parentheses)

$MnOs_3(CO)_{16}H$ (X)	$[Fe_2Mo_2(CO)_{10}Cp_2]^{2-}$ (IX)	$Co_3Fe(CO)_{12}H$ (XII)	$[NiFe_3(CO)_{12}]^{2-}$ (XII)
$MnOs_3(CO)_{13}H_3$ (X, XI)	$[Fe_2W_2(CO)_{10}Cp_2]^{2-}$ (IX)	$Co_3Fe(CO)_{12-n}L_nH$ (XII)	$[NiFe_3(CO)_{12}H]^-$ (XII)
	$FeRu_3(CO)_{13}H_2$ (XI)	$[Co_3Fe(CO)_{11}L]^-$ (XII)	$Ni_2Fe_2(CO)_6Cp_2(C_2R_2)$ (XIII)
	$FeRu_3(CO)_{12}H_4$ (XI)	$[Co_3Fe(CO)_{10}L_2]^-$ (XII)	$[NiCo_3(CO)_{11}]^-$ (XIII)
	$FeOs_3(CO)_{13}H_2$ (XI)	$[Co_3Fe(CO)_{12}]^-$ (XII)	$NiCo_3(CO)_9Cp$ (XII)
	$FeOs_3(CO)_{12}H_4$ (XI)	$[CoFe_3(CO)_{13}]^-$ (XII)	$NiCo_3(CO)_8CpL$ (XII)
	$[Fe_3Co(CO)_{13}]^-$ (XII)	$Co_3Ru(CO)_{12}H$ (X)	
	$FeCo_3(CO)_{12}H$ (XII)	$Co_3Os(CO)_{12}H$ (XII)	
	$FeCo_3(CO)_{12-n}L_nH$ (XII)	$Co_2Os_2(CO)_{12}H_2$ (XI)	
	$[FeCo_3(CO)_{12}]^-$ (XII)	$Co_3Rh(CO)_{12}$ (XII)	
	$[FeCo_3(CO)_{11}L]^-$ (XII)	$Co_2Rh_2(CO)_{12}$ (XII)	
	$[FeCo_3(CO)_{10}L_2]^-$ (XII)	$Co_2Rh_2(CO)_{12-n}L_n$ (XII)	
	$Fe_3Rh(CO)_{11}Cp$ (V)	$Co_2Ir_2(CO)_{12}$ (XII)	
	$Fe_2Rh_2(CO)_8Cp_2$ (V)	$[Co_3Ni(CO)_{11}]^-$ (XIII)	
	$[Fe_3Ni(CO)_{12}]^{2-}$ (XII)	$Co_3Ni(CO)_9Cp$ (XII)	
	$[Fe_3Ni(CO)_{12}H]^-$ (XII)	$Co_3Ni(CO)_8CpL$ (XII)	
	$Fe_2Ni_2(CO)_6Cp_2(C_2R_2)$ (XIII)	$Co_2Pt_2(CO)_8L_2$ (XIII)	
$[Mo_2Fe_2(CO)_{10}Cp_2]^{2-}$ (IX)	$[Ru_3Re(CO)_{16}]^-$ (X)	$Rh_2Fe_2(CO)_8Cp_2$ (V)	$Pd_2Mo_2(CO)_6Cp_2L_2$ (IX)
$Mo_2Pt_2(CO)_6Cp_2L_2$ (IX)	$Ru_2Re_2(CO)_{16}H_2$ (X)	$RhFe_3(CO)_{11}Cp$ (V)	
$Mo_2Pd_2(CO)_6Cp_2L_2$ (IX)	$Ru_3Fe(CO)_{13}H_2$ (XI)	$Rh_2Co_2(CO)_{12}$ (XII)	
	$Ru_3Fe(CO)_{12}H_4$ (XI)	$Rh_2Co_2(CO)_{12-n}L_n$ (XII)	
	$RuCo_3(CO)_{12}H$ (XII)	$RhCo_3(CO)_{12}$ (XII)	
		$Rh_3Ir(CO)_{12}$ (XII)	
		$Rh_2Ir_2(CO)_{12}$ (XII)	

Table 2. (continued)

$[W_2Fe_2(CO)_{10}Cp_2]^{2-}$ (IX)	$[ReRu_3(CO)_{16}]^-$ (X)	$Os_3Mn(CO)_{16}H$ (X)	$Ir_2Co_2(CO)_{12}$ (XII)	$Pt_2Mo_2(CO)_6Cp_2L_2$ (IX)
	$Re_2Ru_2(CO)_{16}H_2$ (X)	$Os_3Mn(CO)_{13}H_3$ (X, XI)	$Ir_2Rh_2(CO)_{12}$ (XII)	$Pt_2Os_2(CO)_8L_2H_2$ (XIII)
	$ReOs_3(CO)_{16}H$ (X)	$Os_3Re(CO)_{16}H$ (X)	$IrRh_3(CO)_{12}$ (XII)	$Pt_2Os_2(CO)_8L_2$ (XIII)
	$ReOs_3(CO)_{15}H$ (X)	$Os_3Re(CO)_{15}H$ (X)		$Pt_2Co_2(CO)_8L_2$ (XIII)
	$ReOs_3(CO)_{13}H_3$ (X, XI)	$Os_3Re(CO)_{13}H_3$ (X, XI)		
		$Os_3Fe(CO)_{13}H_2$ (XI)		
		$Os_3Fe(CO)_{12}H_4$ (XI)		
		$OsCo_3(CO)_{12}H$ (XII)		
		$Os_2Co_2(CO)_{12}H_2$ (XI)		
		$Os_2Pt_2(CO)_8L_2H_2$ (XIII)		
		$Os_2Pt_2(CO)_8L_2$ (XIII)		
		$Os_3(CO)_{10}[Au(PR_3)]$ X (XI)		

Table 3. Illustration of the noble gas rule for tetranuclear cobalt clusters

Compound	Number of electrons donated by the ligands	Experimental number of metal-metal bonds	$\bar{d}_{Co\text{-}Co}$, (Å). (bonding)	$\bar{d}_{Co\text{-}Co}$, (Å). (non-bonding)	Ref.
$Co_4(CO)_9(\mu_2\text{-}CO)_3$	24	6	2.49	–	241, 243)
$Co_4(CO)_6(\mu_2\text{-}CO)_3(C_6H_6)$	24	6	2.455–2.481	–	20)
$Co_4(CO)_8(ffars)_2$	24	6	2.428	–	88)
$Co_4(CO)_8(\mu_2\text{-}CO)_2(C_2Et_2)$	26[1])	5	2.434–2.522	3.547	78)
$Co_4(CO)_8(\mu_2\text{-}CO)_2(\mu_4\text{-}E)_2$					
E = S	28	4	2.48–2.60	–	217)
E = Te	28	4	2.58–2.88	–	217)
E = PPh	28	4	2.519–2.697	–	217)
$Co_4(CO)_4(\mu_3\text{-}SEt)_8$	32	2	2.498	3.312(9)	244)
$Co_4Cp_4(\mu_3\text{-}P)_4$	32	2	2.504	3.630	222)
$Co_4Cp_4(\mu_3\text{-}S)_4$	36	0	–	3.295	223)

[1]) The acetylenic compound C_2Et_2 is assumed to donate 6 electrons by analogy with $Fe_3(CO)_9(C_2Ph_2)$[18a]).

Up to now the main interest in these compounds has been their structural characterization; there is still much to be learned about their methods of synthesis, their chemical reactivity and their catalytic potential.

Generally, the tetranuclear clusters obey the noble gas rule and this is of great assistance in formulating these compounds. Practically, this rule means that each metal atom attains a noble gas configuration by formally allocating a pair of electrons to each metal-metal bond. As discussed previously, this results in a formal metal-metal bond order of 1 [60] and probably corresponds to a fortuitous topological correspondence between the number of bonding delocalized orbitals and the number of edges. Table 3 shows how the noble gas rule is obeyed for tetranuclear cobalt clusters: increasing the number of electrons donated by the ligands results in the progressive breaking of metal-metal bonds.

II. General Trends in M-M and M-CO Energies

The metal carbonyl clusters correspond to situations intermediate between metals and simple mononuclear or binuclear carbonyls. Their existence must be connected either with a delicate thermodynamic balance or with remarkably high activation energies. The last hypothesis is valid for species such as $Rh_6(CO)_{16}$ which is kinetically inert, but in general we are inclined to believe that thermodynamic control is the more significant, especially since reactions, such as that shown in Eq. (1), can be carried out in both directions using mild conditions.

$$2Co_2(CO)_8 \underset{25\,°C,\ 1\ \text{atm CO, iPrOH}}{\overset{60\,°C}{\rightleftarrows}} Co_4(CO)_{12} + 4CO \qquad (1)$$

$$\Delta H° = \Delta H_f^\circ\, Co_4(CO)_{12}(s) + 4\Delta H_f^\circ\, CO(g) - 2\Delta H_f^\circ Co_2(CO)_8(s)$$

$$= -441 - (26.42 \times 4) + (298.9 \times 2) = +51\ \text{kcal mol}^{-1}$$ [68, 98]

(= + 33 kcal mol^{-1} from equilibrium measurements[89])

A condensation process such as that represented in Eq. (1) is strongly endothermic[a]. However, the reaction is possible due to the formation of CO_{gas} which results in a high increase in entropy (the available data indicate that the magnitude of the net term $T\Delta S°$ is ca. 7–8 kcal per mol of CO evolved).

Of particular significance is the fact that very similar reactions, such as:[57, 250]

$$2Rh_2(CO)_8 \underset{-19\,°C,\ 490\ \text{atm}}{\overset{>25\,°C,\ 490\ \text{atm}}{\rightleftarrows}} Rh_4(CO)_{12} + 4CO \qquad (2)$$

a) It is interesting to note that the value obtained from the direct determination of the equilibrium constant, 33 kcal mol^{-1}, seems to be in better agreement with the well known facile interconversion of these two substances.

in which the variation in entropy is practically identical to that in Eq. (1), have very different equilibrium constants. This shows that in this case ΔH° must be much less positive.

A knowledge of these enthalpic terms, and therefore of the relative bond energies, would be expected to considerably clarify many of these fundamental aspects. The data in Table 4 show that, with the main exception of rhenium and osmium, the metal-metal distances in the tetranuclear clusters and in the pure metals are quite similar; this relationship is generally valid for all the polynuclear carbonyls[60]. The metal-metal bond energies in clusters are therefore expected to be of the same order as those in the metallic state: for a close-packed arrangement, these are given by the formula $\bar{D}_{M\text{-}M} = \Delta H_f^\circ M(g)/6$.

Attempts to obtain direct values of the M-M bond energies from mass spectra, or from kinetic data, have provided ambiguous results owing to the possible formation of radicals in excited states[54].

Table 4. Comparison of distances found in metals[229] with average values found in tetranuclear clusters and with values of $\Delta H_f^\circ M(g)/6$[227]

	Mn	Fe	Co	Ni
$\bar{d}_{M\text{-}M}$ (clusters), Å		2.545	2.51	2.51
$d_{M\text{-}M}$ (metal), Å		2.48	2.51	2.49
$\Delta H_f^\circ M(g)/6$, kcal mol^{-1}	(11.35)	16.5	16.8	16.8
		Ru	Rh	
$\bar{d}_{M\text{-}M}$ (clusters), Å		2.79	2.74	
$d_{M\text{-}M}$ (metal), Å		2.65	2.65	
$\Delta H_f^\circ M(g)/6$, kcal mol^{-1}		25.9	22.2	
	Re	Os	Ir	Pt
$\bar{d}_{M\text{-}M}$ (clusters), Å	3.01	2.85	2.71	2.62
$d_{M\text{-}M}$ (metal), Å	2.74	2.675	2.71	2.775
$\Delta H_f^\circ M(g)/6$, kcal mol^{-1}	31	31.5	26.6	22.5

Recently an extensive series of enthalpies of formation of polynuclear carbonyl compounds has been determined, and these values have been divided into the relative bond energies using the following assumptions:[68]

a) The average value of the metal-carbon bond energy $\bar{D}_{M\text{-}CO}$ is similar for terminal and bridging carbonyl groups.

b) The average values of the metal-carbonyl bond energy $\bar{D}_{M\text{-}CO}$, is independent of nuclearity.

c) The average value of $\bar{D}_{M\text{-}M}$ is also independent of nuclearity.

d) The following empirical relationship has been found for iron and cobalt carbonyls:

$$\bar{D}_{M\text{-}M} \simeq 0.68\bar{D}_{M\text{-}CO}$$

and it has been assumed that this relationship can be applied to the other transition metals.

The values obtained using these approximations are reported in Table 5.

Table 5. Average value of the M-M and M-CO bond energies, according to Connor[68] (in kcal/mol^{-1})

	Cr	Mn	Fe	Co	Ni
$\bar{D}_{M\text{-}CO}$	26	24	28.1	32.5	35.1
$\bar{D}_{M\text{-}M}$		16	19.2	22	
	Mo		Ru	Rh	
$\bar{D}_{M\text{-}CO}$	36		41.2	39	
$\bar{D}_{M\text{-}M}$			28	26.5	
	W	Re	Os	Ir	
$\bar{D}_{M\text{-}CO}$	42	44.8	45.8	45.3	
$\bar{D}_{M\text{-}M}$		30.5	31.1	31	

Actually a comparison both of M-M and M-CO distances does not show large variations due to changes in the nuclearity of the clusters, and moreover the available information about the equilibria involving bridged and unbridged species[37, 188, 203] indicates that there are only minor overall differences in the relative bond energies. Unfortunately, although these average bond energies seem at first sight reasonable, their use in reactions, in which only a limited number of metal-metal bonds is involved, is probably unrealistic due to the high π-acidity of carbon monoxide and to the possible variation of the single $D_{C\text{-}O}$ terms. The situation is often particularly complicated by the presence of negative charges and the following series of reactions can be cited in order to show that minor differences in bond energies, which in this case are due to increased back donation from negative charges, are responsible for the variation in stability[73, 111, 185].

$$Rh_4(CO)_{12} + CO \xrightarrow[1\ \text{atm, THF}]{25\ °C} \text{no reaction (but rapid exchange with } ^{13}CO) \qquad (3)$$

$$[Rh_4(CO)_{11}(COOMe)]^- + CO \xrightarrow[1\ \text{atm, THF}]{25\ °C} \text{no reaction (but rapid exchange with } ^{13}CO) \qquad (4)$$

$$2[Rh_4(CO)_{11}]^{2-} + CO \xrightarrow[1\ \text{atm, THF}]{25\ °C} 2[Rh(CO)_4]^- + [Rh_6(CO)_{15}]^{2-} \qquad (5)$$

Finally it has been pointed out that, within the same transition period, the trend in $\bar{D}_{M\text{-}CO}$ is obscured by the different promotion energies to the valence state and a trend different to that in Table 5 has been calculated for the first period (Cr = 54.2; Fe = 58.7; Ni = 45.6 kcal mol^{-1})[225]. A similar effect probably occurs on descending a subgroup.

III. General Methods of Synthesis

The synthesis of tetranuclear clusters is generally carried out using one of the following general methods, which we will discuss only briefly because they have already been described in previous reviews[54, 60], but it is appropriate to emphasize that we are not yet in a position to forecast the result of any of these syntheses.

A. Condensation Between Coordinatively Unsaturated Species

The most important synthetic method is through coordinatively unsaturated species, generated by thermal, photochemical or chemical methods. An example is reaction (1) which gives $Co_4(CO)_{12}$, where kinetic studies are in agreement with the intermediate formation of $Co_2(CO)_7$ [235, 236].

This type of reaction often takes place with contemporary elimination of a stable species, as for instance in the sequence:[237, 238]

$$2Co_3(CO)_3Cp_3 \xrightarrow[130\,^{\circ}C]{\Delta} 2Co(CO)_2Cp + 2[Co_2(CO)Cp_2] \longrightarrow Co_4(CO)_2Cp_4 \quad (6)$$

although there is much ambiguity about polynuclear coordinatively unsaturated species due to the possible formation of multiple metal-metal bonds [for instance $Co_2(CO)_2Cp_2$ [221) $Cr_2(CO)_4(\eta^5\text{-}C_5Me_5)_2$ [211], $Os_3(CO)_{10}H_2$ [187], $Re_4(CO)_{12}H_4$ [218]]. An example of coordinatively unsaturated species generated in a chemical way is believed to be represented by the following reactions:

$$2Co_2(CO)_8 \xrightarrow[\text{hexane}]{Rh_2(CO)_4Cl_2} Co_2Rh_2(CO)_{12} + 2CoCl_2 + 8CO \quad \text{(yield 80–90\%)}\ [184]$$

$$2Co_2(CO)_8 \xrightarrow[\text{hexane}]{Rh_2(PF_3)_4Cl_2} Co_2Rh_2(CO)_8(PF_3)_4 + 2CoCl_2 + 8CO\ [128] \quad (7)$$

The photochemical generation of coordinatively unsaturated species offers great potential in this area but unfortunately has not yet been paid sufficient attention.

B. Redox Condensation

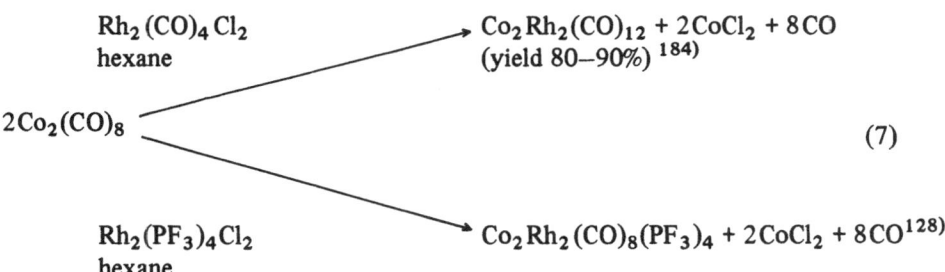

This method consists of the condensation of two species in different oxidation states, and is particularly important in the chemistry of high nuclearity carbonyl clusters[60]. These reactions often take place in very mild conditions showing that their redox character is associated with low activation energies.

Some examples of redox condensations are:

$$[Fe_3(CO)_{11}]^{2-} + Fe(CO)_5 \xrightarrow[Py]{85\,^\circ C} [Fe_4(CO)_{13}]^{2-} + 3CO \text{ (Ref.}^{127)}\text{)} \quad (8)$$

$$[Fe_3(CO)_{11}]^{2-} + Ni(CO)_4 \xrightarrow[THF]{25\,^\circ C} [NiFe_3(CO)_{12}]^{2-} + 3\,CO \text{ (Ref.}^{172)}\text{)} \quad (9)$$

$$2Fe_3(CO)_{12} + [Co(CO)_4]^- \xrightarrow[THF]{25\,^\circ C} [Fe_3Co(CO)_{13}]^- + 3Fe(CO)_5 \text{ (Ref.}^{61)}\text{)} \quad (10)$$

In order to emphasize the limitations of this method, the above results should be compared with the simple redistribution which takes place in the following reaction:[8]

$$Fe_3(CO)_{12} + [Mn(CO)_5]^- \xrightarrow[diglyme]{} [MnFe_2(CO)_{12}]^- + Fe(CO)_5 \quad (11)$$

C. Condensation Induced by Particular Ligands

Some post-transition elements (or the corresponding radicals) containing 3 or more electrons in their valence shell are able to assist the formation of clusters by bonding to several metal atoms. Typical examples of this behaviour are the extraordinarily easy syntheses of large series of compounds such as $Co_3(CO)_9(\mu_3\text{-}E)$ (E = Al, CR, CX, GeR, P, As, PS, S, Se, PR, SR)[207, 209] and $Fe_3(CO)_9(\mu_3\text{-}E)_2$ (E = S, Se, Te, NR, PR). This type of stabilization is usually found in trinuclear clusters although a few examples in tetranuclear clusters are known, for instance:

$$Co_2(CO)_8 \xrightarrow[heptane]{S_8} Co_4(CO)_{10}(\mu_4\text{-}S)_2 \text{ (yield ca. 33\%)}^{179)}$$

$$Co_2(CO)_8 \xrightarrow[Ph_2Te]{benzene} Co_4(CO)_{10}(\mu_4\text{-}Te)_2 \; ^{125,\,217)} \quad (12)$$

$$Co_2(CO)_8 \xrightarrow{Zn\,+\,PhPCl_2} Co_4(CO)_{10}(\mu_4\text{-}PPh)_2 \; ^{217)}$$

As shown in Fig. 1, the two hetero-atoms are above and below the centre of a rectangle of cobalt atoms. As a first approximation the coordination around the hetero-atoms corresponds to a pyramid in which the axial position is occupied by a lone-pair of electrons (or by the Ph group) while the four equatorial positions formally correspond to the four electrons donated by the hetero-atom.

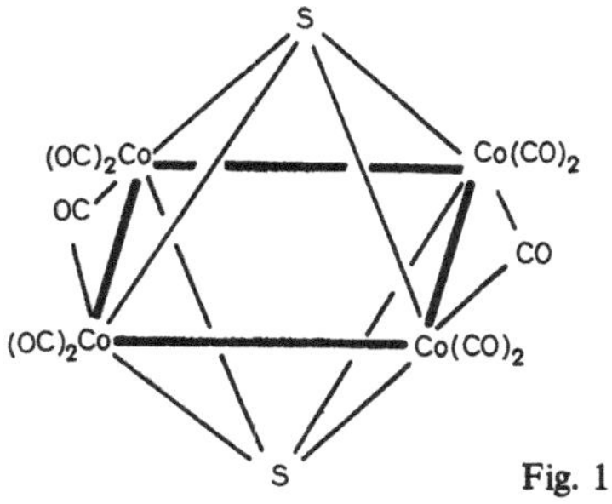

Fig. 1. Schematic structure of $Co_4(CO)_8(\mu_2\text{-}CO)_2(\mu_4\text{-}S)_2$ [217]

A much more subtle case of stabilisation due to the presence of face-bridging ligands is found in the two tetrahedral clusters $Co_4Cp_4(\mu_3\text{-}H)_4$ (dec. ca. 300 °C)[134] and $Ni_4Cp_4(\mu_3\text{-}H)_3$ (dec. ca. 320 °C)[133]. The structure of the last compound is shown in Fig. 2.

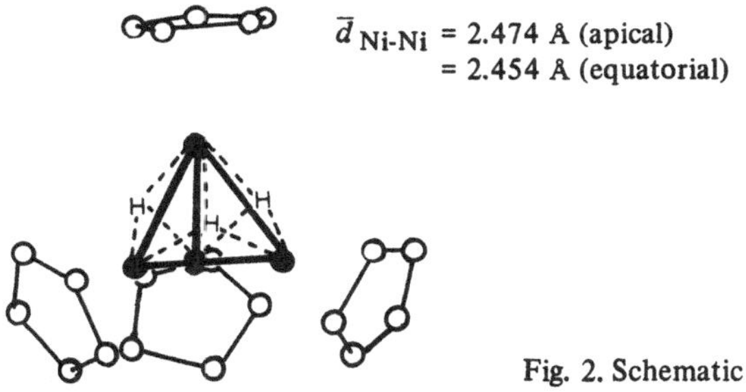

Fig. 2. Schematic structure of $Ni_4Cp_4(\mu_3\text{-}H)_3$ [133]

This nickel compound is particularly interesting because, in spite of its paramagnetism (3 unpaired electrons), it shows no tendency to lose hydrogen on heating to give the expected diamagnetic cluster Ni_4Cp_4 [198]. The two compounds have been prepared very simply by reacting AlH_3 in THF at 20 °C with $Co_2Cp_2(NO)_2$ [198] and $NiCp(NO)$[199] respectively.

Stabilization due to relief of steric crowding is probably involved in the following remarkable reactions which take place at atmospheric pressure:

$$Re_2(CO)_{10} \xrightarrow[H_2]{150\,°C} Re_3(CO)_{12}H_3 \xrightarrow[H_2,\ 50h]{150\,°C} Re_4(CO)_{12}H_4 \quad \text{(yield 50–60\%)}\ ^{148)} \tag{13}$$

$$Ru_3(CO)_{12} \xrightarrow[H_2]{90\,°C,\ octane} Ru_4(CO)_{12}H_4 \quad \text{(yield 88\%)}\ ^{163)} \tag{14}$$

$$Os_3(CO)_{12} \xrightarrow[H_2]{110\,°C,\ octane} Os_3(CO)_{12}H_2 \xrightarrow[H_2]{110\,°C,\ octane} Os_4(CO)_{12}H_4 \quad \text{(yield 29\%)}\ ^{163)} \tag{15}$$

Table 6. Bond distances (Å) in tetranuclear metal carbonyl closed clusters

Iron, cobalt and nickel	Number of M-M bonds	Idealized symmetry	M-M in mixed clusters	$\bar{d}_{M\text{-}M}$	$\bar{d}_{M\text{-}M}$ (H-bridged)	$\bar{d}_{M\text{-}CO}$ terminal	$\bar{d}_{M\text{-}CO}$ edge-bridging	$\bar{d}_{M\text{-}CO}$ face-bridging	Ref.
$[Fe_4(CO)_{13}H]^-$	5	C_s	–	2.63	–	1.75	–	–	178)
$[Fe_4(CO)_{13}]^{2-}$	6	C_{3v}	–	2.545	–	1.72	–	2.00	85)
$Fe_4(CO)_4Cp_4$	6	T_d	–	2.520(1)	–	–	–	1.986(6)	201)
$[Fe_4(CO)_4Cp_4]^+$	6	D_{2d}	–	2.484(2)	–	–	–	1.979(12)	234)
$Fe_2Rh_2(CO)_8Cp_2$	6	C_{2v}	Fe-Fe	2.539(7)	–	1.72	–	–	65)
$Fe_3Rh(CO)_{11}Cp$	6	C_s	Fe-Fe	2.577	–	1.82	–	–	64)
$Co_4(CO)_{12}$	6	C_{3v}	–	2.49	–	1.83	2.04	–	43, 243)
$Co_4(CO)_9$(arene)[1]	6	C_{3v}	–	2.47	–	1.785(13)	1.928	–	20)
$Co_4(CO)_{10}(C_2Et_2)$	5	C_2	–	2.458	–	1.73	1.84–1.97	–	78)
$Co_4(CO)_{10}(S)_2$	4	D_{2h}	–	2.54	–	–	–	–	217)
$Co_4(CO)_{10}(Te)_2$	4	D_{2h}	–	2.73	–	–	–	–	217)
$Co_4(CO)_{10}(PPh)_2$	4	D_{2h}	–	2.608	–	1.777(8)	1.928(8)	–	217)
						1.824(9)	1.959(9)		
$Co_4(CO)_8$(ffars)$_2$	6	C_2	–	2.428	–	1.75	–	–	88)
$Co_2Ir_2(CO)_{12}$	6	C_{2v}	Co-Co	2.644	–	1.78	–	–	4)
$Co_3Fe(CO)_9[P(OMe)_3]_3H$	6	C_{3v}	Co-Co	2.488(12)	–	–	–	–	132)
$Co_2Pt_2(CO)_3(PPh_3)_2$	5	C_1	Co-Co	2.498(3)	–	–	–	–	95)
$Ni_4(CO)_6[P(CH_2CH_2CN)_3]_4$	6	T_d	–	2.508(4)	–	–	1.890(3)	–	17)
$Ni_4(CO)_4(CF_3C_2CF_3)_3$	6	C_{3v}	–	2.524	–	1.81	–	–	79)
Ruthenium and rhodium									
$Ru_4(CO)_{13}H_2$	6	C_s	–	2.783	2.930	1.89(10)	1.94–2.40	–	253)
$Ru_4(CO)_{11}(C_8H_{10})$[2]	5	C_s	–	2.755	–	–	–	–	186)
$Ru_4(CO)_{10}(C_{12}H_{16})$[3]	5	?	–	2.797	–	1.87	–	–	16)
$Ru_4(CO)_9(C_{13}H_{14})$[4]	6	C_s	–	2.82	–	1.861(15)	–	–	63)
$Ru_3Fe(CO)_{13}H_2$	6	C_s	Ru-Ru	2.796	2.904	1.82(8)	2.28(4)	–	100)
$Rh_4(CO)_{12}$	6	C_{3v}	–	2.73	–	1.96	1.99	–	243)

$[Rh_4(CO)_{11}]^{2-}$	6	C_{3v}	–	2.75	–	1.80	2.11	–	6)
$Rh_4(CO)_8(DME)_2$	6	C_1	–	2.711	–	1.91	2.09	–	43)
$Rh_2Fe_2(CO)_8Cp_2$	6	C_{2v}	Rh-Rh	2.683	–	–	1.97(4)	–	65)
Rhenium, osmium, iridium and platinum									
$Re_4(CO)_{12}H_4$	6–8	T_d	–	2.913(8)	–	1.90(2)	–	–	251a)
$[Re_4(CO)_{16}]^{2-}$	5	D_{2h}	–	2.99	–	–	–	–	11)
$[Re_4(CO)_{15}H_4]^{2-}$	4	C_1	–	3.023	3.23	1.84	–	–	5)
$[Re_4(CO)_{12}H_6]^{2-}$	6	T_d	–	–	3.160(2)	–	–	–	145)
$Re_4(CO)_{12}[\mu_3\text{-}InRe(CO)_5]_4$	6	T_d	–	3.028(5)	–	–	–	–	108)
$Os_3Au(CO)_{10}(PPh_3)X$[5]	5	C_s	Os-Os	2.85	–	–	–	–	26)
$Ir_4(CO)_{12}$	6	T_d	–	2.68	–	1.90	–	–	240, 242)
$[Ir_4(CO)_{10}H_2]^{2-}$	6	C_s	–	2.76	–	1.80	2.12–1.90	–	102)
$Ir_4(CO)_{12-n}(PPh_3)_n$[6]	6	C_1-C_s		2.73	–	1.85	2.1		3)
$Pt_4(CO)_5(PMe_2Ph)_4$	5	C_{2v}	–	2.616	–	–	–	–	239)

1) Arene = benzene, mesitylene.
2) C_8H_{10} = cyclo-octene-5-yne.
3) $C_{12}H_{16}$ = cyclododecatetraene.
4) $C_{13}H_{14}$ = 4,6,8-trimethylazulene.
5) X = Cl and Br.
6) n = 2 and 3.

IV. Methods of Separation

The study of the properties and applications of clusters requires them to be accessible on a preparative scale and this is often conditioned by the ease with which they can be separated from other reaction products. Because their molecular weights are not too high, several tetranuclear clusters are volatile and/or soluble in organic solvents, and therefore the first methods of separation to be considered are sublimation under high vacuum (if the compound is thermally stable) and crystallization. When the solubility is low, continuous extraction with low boiling solvents can be tried[50]. An important example illustrating the difficulties in this area is $Ir_4(CO)_{12}$, which is only accessible on a preparative scale in sufficient and reproducible purity using particular techniques such as a continuous hot solvent extraction.

The chromatographic technique is important especially in initial separations of mixtures and in the final purification of compounds. However, owing to the limited amounts of substances which can be produced using such techniques, it would be desirable that, at some later stage, serious efforts aimed at developing better preparative routes to these clusters should be made.

The chemistry of the tetranuclear carbonyl anions is not very well developed but, in agreement with the behaviour of anions which contain 3–4 metal atoms per negative charge, the monoanion $[FeCo_3(CO)_{12}]^-$ has been separated in good yield as the potassium salt by simple addition of excess potassium acetate or bromide in aqueous solution[52]. The salts of the monoanions with large cations such as PPN^+ are often soluble in organic solvents ($PPN[Rh_4(CO)_{11}(COOMe)]$ is even soluble in toluene[185]) and sometimes can be extracted selectively.

V. Structural Data in the Solid State

A. X-ray

The available structural data on the tetranuclear carbonyl clusters are summarized in Table 6.

The most common geometrical arrangement of metals in tetrametal clusters is the tetrahedron, a polyhedron which represents the simplest case of a 3-dimensional cluster. It is convenient to start the discussion by considering the basic structures of the dodecacarbonyls and then to consider the effects produced by increasing and decreasing the number of carbonyl groups around the tetrahedral unit of metals.

The high symmetry T_d of the tetrahedron is retained in the structure of $Ir_4(CO)_{12}$ represented in Fig. 3.

In the right part of Fig. 3 is the cube octahedron corresponding to the envelope of carbonyl groups. On the basis of the experimental data ($d_{Ir\text{-}Ir}$ = 2.68 Å [242] and $d_{Ir\text{-}CO}$ = 1.90 Å [240]) the C···C contacts have been calculated to be about 3.2 Å, a

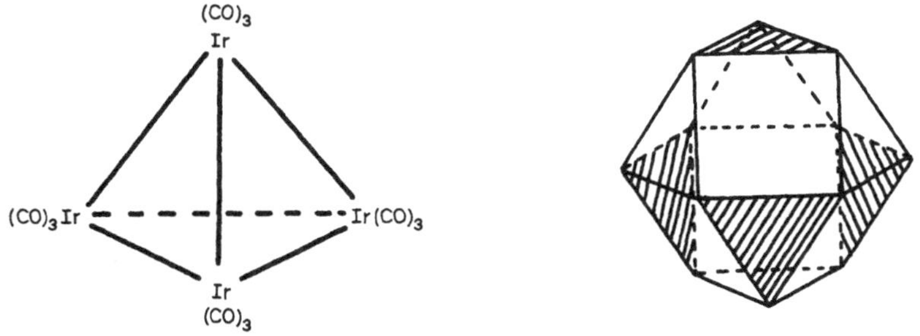

Fig. 3. Schematic structure of $Ir_4(CO)_{12}$, (T_d), and the polyhedron defined by the carbonyl groups[240, 242]

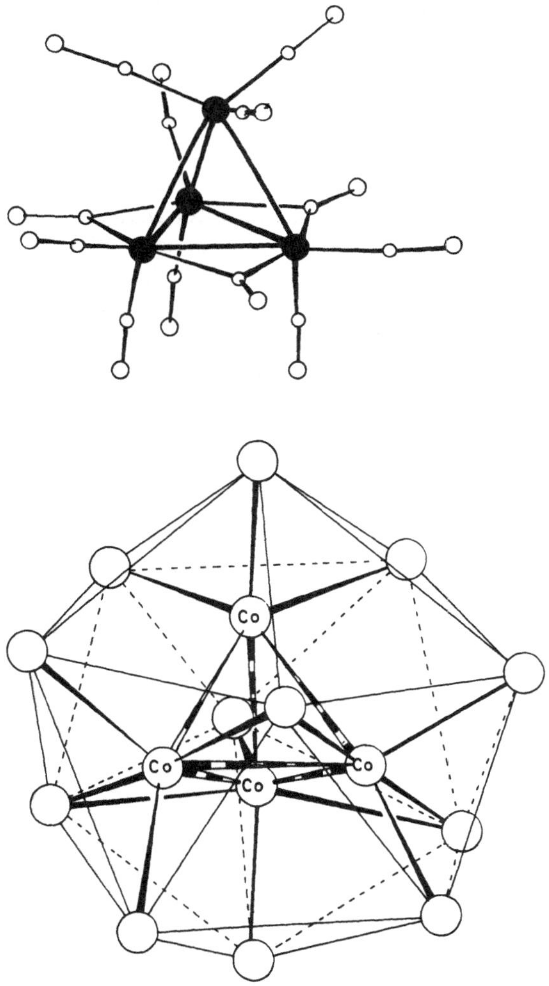

$\bar{d}_{Co\text{-}Co} = 2.492$ Å
$\bar{d}_{Co\text{-}CO_t} = 1.834$ Å
$\bar{d}_{Co\text{-}CO_b} = 2.043$ Å

$\bar{d}_{Rh\text{-}Rh} = 2.73$ Å
$\bar{d}_{Rh\text{-}CO_t} = 1.96$ Å
$\bar{d}_{Rh\text{-}CO_b} = 1.99$ Å

Fig. 4. Schematic structure of $Co_4(CO)_9(\mu_2\text{-}CO)_3$, ($C_{3v}$), and the polyhedron defined by the carbonyl groups[43, 85, 243]

distance significantly greater than the observed Van der Waal's contacts of ca. 2.5–2.6 Å[b].

The tetrahedral structure of $Co_4(CO)_{12}$[243], which is also common to $Rh_4(CO)_{12}$[243] and to $Co_2Ir_2(CO)_{12}$[4], is shown in Fig. 4.

In the cobalt case, the C···C contacts in the icosahedron have been calculated to be about 3.0 Å, which is still larger than the Van der Waal's contacts[c]. Moreover, because $d(M–CO_{bridging})$ is longer than $d(M–CO_{terminal})$, there is some further relief of steric-crowding due to the carbon atoms not being equi-distant from the metal atoms and this results in a larger surface of the polyhedron. The low frequency of the bridging carbonyl groups, corresponding to progressive transformation from C≡O to C=O, shows that back-donation towards these bridging groups is greater and their formation is also related to the different tendency of the metals to lose negative charge[54, 74]. A compromise situation between steric and electronic effects is generally believed to be responsible for bridge formation[54].

Because it is generally not possible to distribute in a perfectly symmetrical way the coordination positions around the metal skeleton, the surface at the level of the carbon atoms is not completely occupied and steric compression is frequently present on some part of such a surface. Actually it seems probable that owing to the increasing number of carbonyl groups, the tetrahedral clusters of the transition metals which precede group VIII are sterically destabilized. Of the three tetranuclear rhenium clusters of known structures, $[Re_4(CO)_{12}H_6]^{2-}$ (Ref.[145]), $[Re_4(CO)_{15}H_4]^{2-}$ (Ref.[5]), and $[Re_4(CO)_{16}]^{2-}$(Ref.[11]), only the first adopts a tetrahedral structure similar to $Ir_4(CO)_{12}$ in which the six H atoms are believed to be bonded along the 6 tetrahedral edges [$d_{Re\text{-}Re}$ = 3.160 (7) Å][145]. The other two clusters have the structures shown in Fig. 5.

The existence of $Re_4(CO)_{12}H_4$, which contains only 56 valence electrons, and which corresponds to a tetrahedron having rhenium-rhenium multiple bonds[218], should be considered from the same point of view. The structure of this cluster has been confirmed very recently[251a]: the average Re-Re distance of 2.913(8) Å is significantly short, the hydrogen atoms triply bridge the tetrahedral faces, and the geometrical distribution of the twelve carbonyls corresponds to a truncated tetrahedron.

The presence of long Re-In distances is probably responsible for the high stability of the central metallic tetrahedron in the cluster $Re_4(CO)_{12}[\mu_3\text{-InRe}(CO)_5]_4$[108] shown in Fig. 6.

It has been reported that the steric requirements of the (η^5-C_5H_5) group are approximately equal to that of 2.5 carbonyl groups[159], although again the homogeneous distribution of cyclopentadienyl ligands around a cluster is often limited. Actually, the compound $Fe_4(CO)_4Cp_4$, which has the structure shown in Fig. 7, formally corresponds to the unknown $Fe_4(CO)_{14}$.

---

[b] Van der Waal's contacts between groups bonded to the same centre are known to depend on the relative angle between these groups; the minimum observed values of 2.5–2.6 Å refer to the common situation in which OC–$\hat{M}$–CO angles of about 90–110 °C are present.

[c] An *ad hoc* model based on Van der Waal's contacts of 3.02 Å and on closepacking of carbonyl groups has been claimed to rationalise this structural chemistry, but it has produced mainly unrealistic results such as, for instance, a cube octahedral arrangement of carbonyl groups in $Rh_4(CO)_{12}$, $Rh_3Ir(CO)_{12}$ and $Rh_2Ir_2(CO)_{12}$[144].

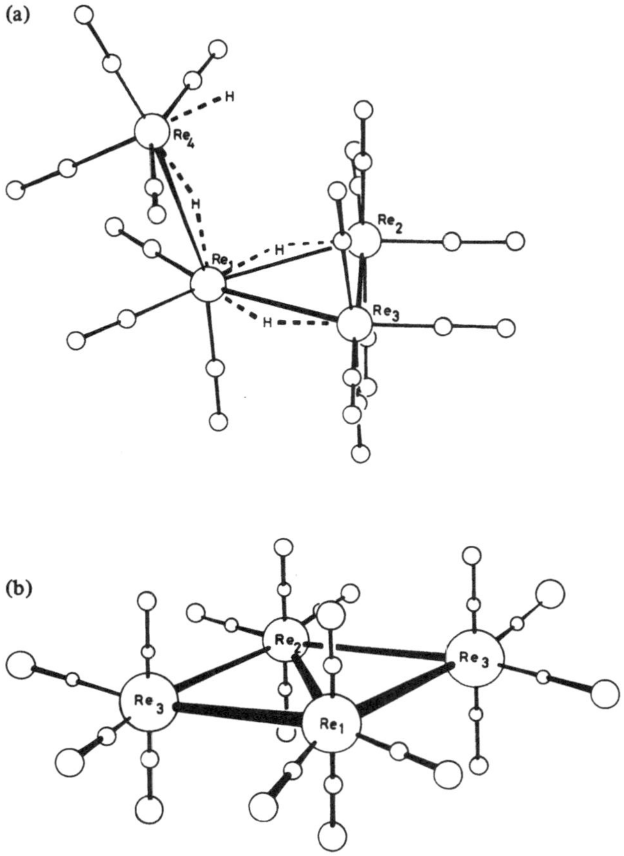

$d_{Re_1\text{-}Re_4}$ = 3.288 Å
$d_{Re_1\text{-}Re_2}$ = 3.211 Å
$d_{Re_1\text{-}Re_3}$ = 3.192 Å
$d_{Re_2\text{-}Re_3}$ = 3.023 Å

$d_{Re_1\text{-}Re_2}$ = 2.96 Å
$d_{Re_2\text{-}Re_3}$ = 3.00 Å

Fig. 5. Schematic structures of the anions $[Re_4(CO)_{15}H_4]^{2-}$ (C_1) and $[Re_4(CO)_{16}]^{2-}$ (D_{2h})[5, 11]

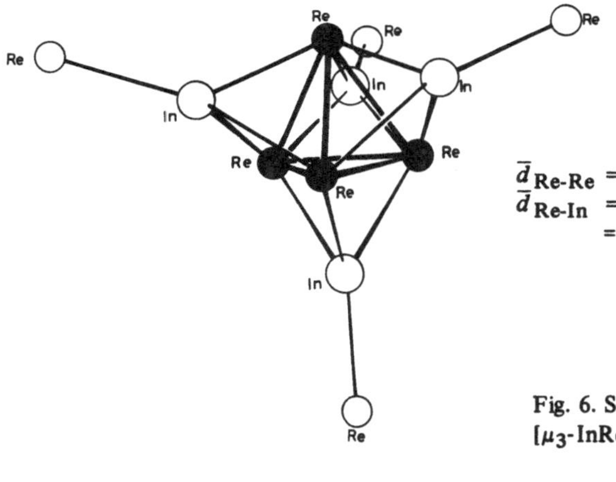

$\bar{d}_{Re\text{-}Re}$ = 3.028(5) Å
$\bar{d}_{Re\text{-}In}$ = 2.818(7) Å (to the Re_4 tetrahedron)
= 2.735(7) Å (to the external Re)

Fig. 6. Schematic structure of $Re_4(CO)_{12}$-$[\mu_3\text{-}InRe(CO)_5]_4$, ($T_d$)[108]

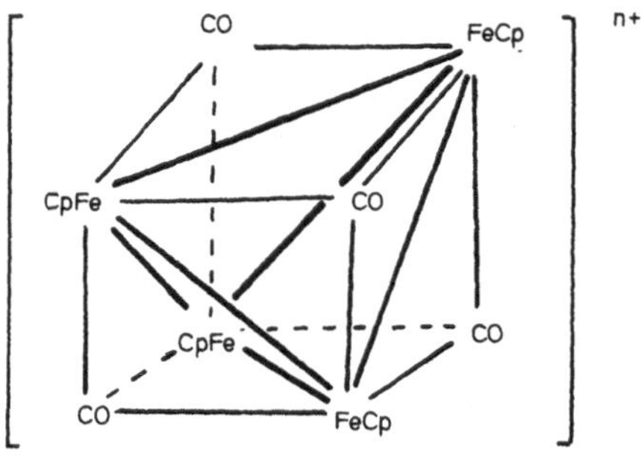

n	$d_{Fe\text{-}Fe}$, Å	$\bar{d}_{Fe\text{-}CO_b}$, Å
$0(T_d)$	2.520(1)	1.986(6)
$1(D_{2d})$	2.484(2)	1.979(12)

Fig. 7. Schematic structure of $[Fe_4(\mu_3\text{-}CO)_4Cp_4]^n$ (n = 0,[201] + 1 [234])

(a)

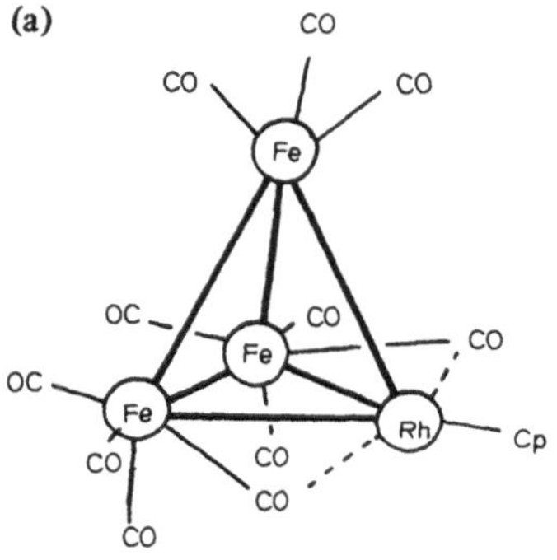

$\bar{d}_{Fe\text{-}Fe}$ = 2.577 Å $\bar{d}_{Fe\text{-}Rh}$ = 2.593 Å
$\bar{d}_{Fe\text{-}CO_t}$ = 1.822 (apical), 1,825 (basal) Å

(b)

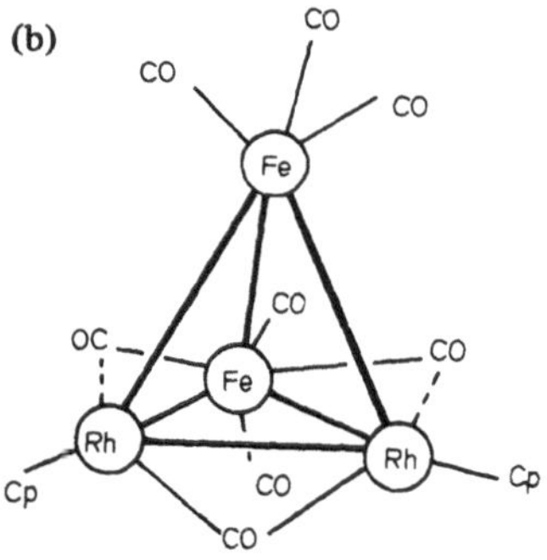

$d_{Fe\text{-}Fe}$ = 2.539(7) Å $d_{Rh\text{-}Rh}$ = 2.683(3) Å
$\bar{d}_{Rh\text{-}Fe}$ = 2.570(5) and 2.598(5) Å
$\bar{d}_{Fe\text{-}CO_t}$ = 1.69 (apical), 1.79 (basal) Å

Fig. 8. Schematic structures of
(a) $Fe_3Rh(CO)_9(\mu_2\text{-}CO)_2Cp$, (C_s) and
(b) $Fe_2Rh_2(CO)_5(\mu_2\text{-}CO)_3Cp_2$, (C_{2v})[64, 65]

In the corresponding cation $[Fe_4Cp_4(CO)_4]^+$, there is a slight deformation towards D_{2d} symmetry although all the metal-metal distances have decreased to an average value of 2.484 Å in agreement with the formation of some multiple bonding between the metal atoms.

In the tetrahedra, $Fe_3Rh(CO)_{11}Cp$, and $Fe_2Rh_2(CO)_8Cp_2$, shown in Fig. 8 there is a marked crowding of the ligands in the basal plane[64, 65]. In both cases, the apical iron atoms formally have only 17 electrons whereas the electronic situation of the iron atoms in the basal plane is ill-defined due to the presence of asymmetric bridges. A comparison of the $Fe\text{-}CO_{terminal}$ distances indicates that formation of bridges, especially in the second case, leaves less electronic density in the basal plane which results in longer Fe-$CO_{terminal}$ distances.

The structure in Fig. 8a is also similar to the structures of the dihydrides, $MRu_3(CO)_{13}H_2$ (M = Fe[100] or Ru[253]), shown in Fig. 9. The six carbonyl groups present in the basal plane of the dihydrides are inhomogeneously distributed and this gives rise to C···C contacts of the order of 2.5 – 2.6 Å which results in considerable steric compression in this region.

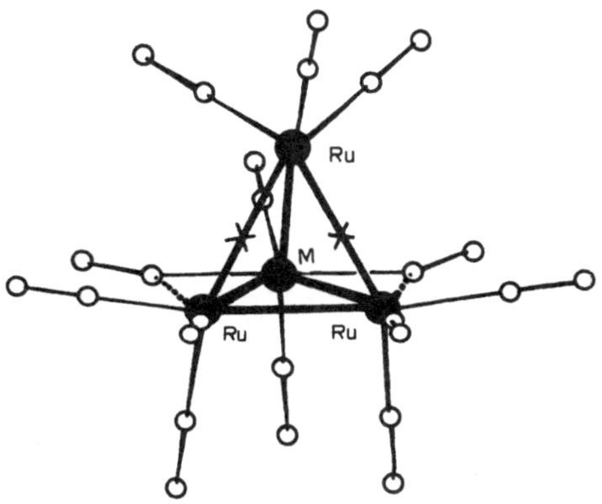

M = Fe		Ru
$\bar{d}_{RuHRu}$	= 2.904 Å	2.930Å
$\bar{d}_{Ru\text{-}Ru}$	= 2.796 Å	2.783 Å
$\bar{d}_{Fe\text{-}Ru}$	= 2.662 Å	
$\bar{d}_{Ru\text{-}CO}$	= 1.82 Å	1.89 Å
$\bar{d}_{Fe\text{-}CO}$	= 1.72 Å	

X = probable position of H's.

Fig. 9. Schematic structure of $MRu_3(CO)_{11}$-$(\mu_2\text{-}CO)_2H_2$, (M=Fe or Ru), (C_s)[100, 253]

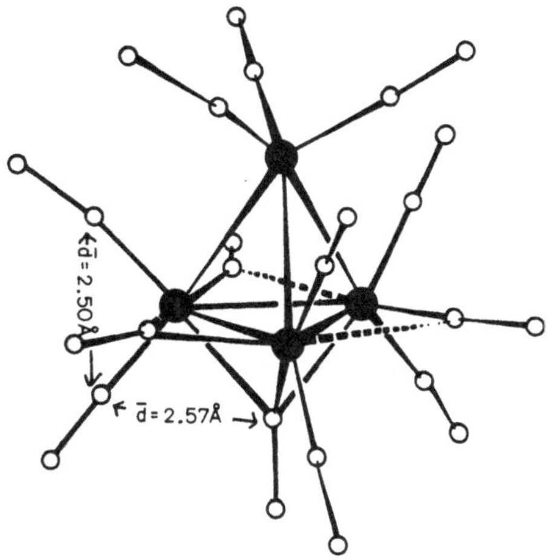

$\bar{d}_{Fe\text{-}Fe}$ = 2.59(3) Å (apical)
= 2.50(3) Å (basal)
$\bar{d}_{Fe\text{-}CO_t}$ = 1.72 Å
$\bar{d}_{Fe\text{-}CO_b}$ = 2.00 Å (μ_3)

(shortest contacts 2.46 – 2.61 Å)

Fig. 10. Schematic structure of the anion $[Fe_4(CO)_9(\mu_2\text{-}CO)_3(\mu_3\text{-}CO)]^{2-}$, (C_{3v})[85]

A similar steric compression is also present in the anion $[Fe_4(CO)_{13}]^{2-}$ in which the thirteenth carbonyl group is now under the basal plane, (Fig. 10). Although the average C···C contact value is 2.76 Å, some of these contacts, particularly with the face-bridging carbonyl group, reach the limiting value of 2.5 – 2.6 Å. In this case, the 3 edge-bridging carbonyls in the basal plane are extremely asymmetric.

The basic distribution of ligands present in $Co_4(CO)_{12}$ is also maintained in the recently characterized $[Ir_4(CO)_{10}H_2]^{2-}$ dianion[102], shown in Fig. 10a. The two Ir-Ir distances (1–4 and 1–3) *trans* to the assumed positions of the hydrogen atoms are particularly long (2.80 Å).

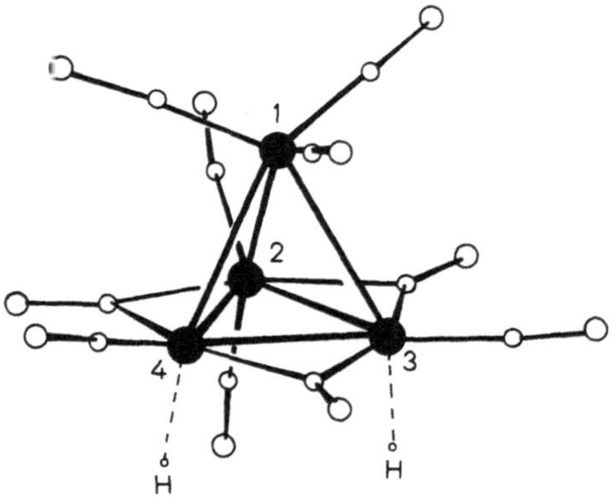

$\bar{d}_{Ir\text{-}Ir}$ = 2.76 Å
$\bar{d}_{Ir\text{-}CO_t}$ = 1.80 Å
$\bar{d}_{Ir\text{-}CO_b}$ = 1.90 (3–4)
2.12 (2–3 and 2–4) Å

Fig. 10a. Schematic structure of the dianion $[Ir_4(CO)_{10}H_2]^{2-}$, (C_s)[102]

Decreasing the number of ligands around the tetrahedral cluster results in large variations in the way in which they are distributed. Figure 11 shows the structure of the anion $[Rh_4(CO)_{11}]^{2-}$[6]. In this case, the presence of seven bridging carbonyl groups is in keeping with the double negative charge, although steric effects must also be significant (compare with the four bridging carbonyls present in $[Fe_4(CO)_{13}]^{2-}$). Due to the uneven distribution of carbonyl groups, the Rh_b atoms formally attain a 17 electron configuration and probably most of the negative charge is associated with these two metal atoms accounting for the long Rh_b-Rh_b distance (2.99 Å). The differences between the Rh-$CO_{terminal}$ distances [1.765(2) Å for Rh_b and 1.83(2) Å for Rh_a] also agrees with the different number of bridging carbonyl groups present on these two types of atoms.

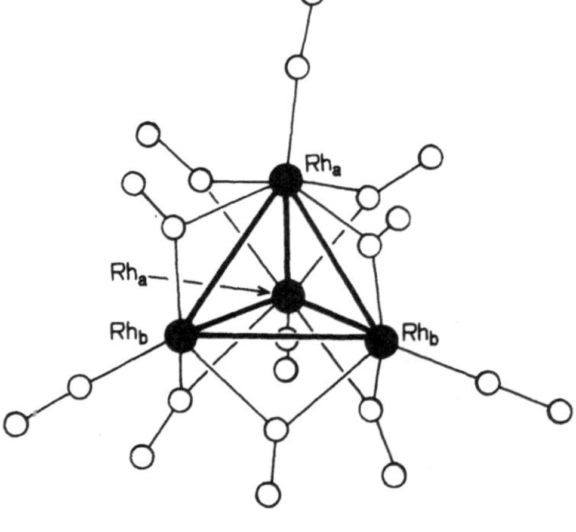

$\bar{d}_{Rh\text{-}Rh}$ = 2.75 Å

Fig. 11. Schematic structure of the anion $[Rh_4(CO)_4(\mu_2\text{-}CO)_7]^{2-}$, (C_{2v})[6]

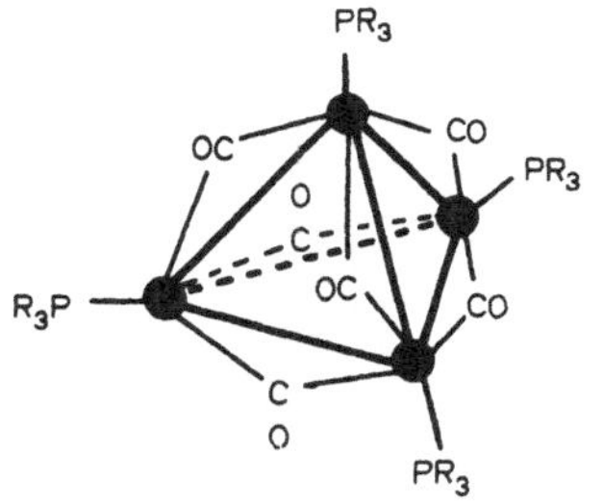

$\bar{d}_{Ni\text{-}Ni}$ = 2.508(4) Å
$\bar{d}_{Ni\text{-}CO}$ = 1.890(3) Å
$\bar{d}_{C=O}$ = 1.20(2) Å

Fig. 12. Schematic structure of $Ni_4(\mu_2\text{-}CO)_6[P(CH_2CH_2CN)_3]_4$, (T_d)[17)]

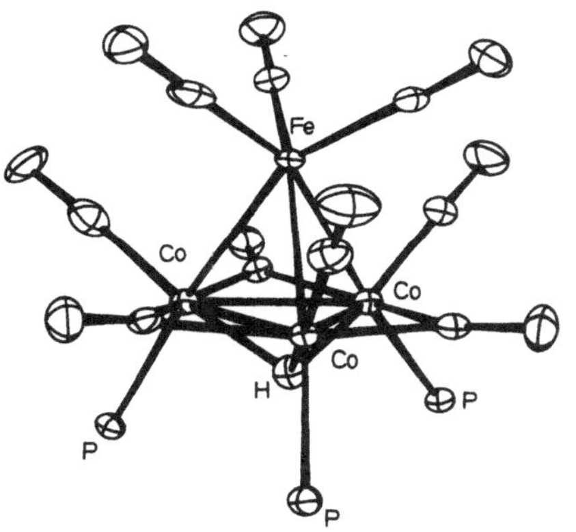

$\bar{d}_{Fe\text{-}Co}$ = 2.560(2) Å
$\bar{d}_{Co\text{-}Co}$ = 2.488(12) Å
$\bar{d}_{Co\text{-}H}$ = 1.63(15) Å

Fig. 13. Schematic structure of $FeCo_3(CO)_6(\mu_2\text{-}CO)_3\text{-}[P(OMe)_3]_3(\mu_3\text{-}H)$, (C_s)[132)]

In the cluster $Ni_4(\mu_2\text{-}CO)_6[P(CH_2CH_2CN)_3]_4$, the presence of just 10 ligands gives rise to a most symmetrical tetrahedral structure, which has all edges occupied by bridging carbonyl groups, (Fig. 12)[17)].

The steric requirement of a tertiary phosphine is difficult to compare with that of CO: not only does it depend on the groups on phosphorus[233)] but, because of the longer M-P distance, there is some relief of steric crowding at the carbon level. Steric effects due to the relative requirements of the groups bonded to phosphorus have generally been found only in tri- and/or tetrasubstituted compounds depending on the relative positions of the phosphorus ligands. Furthermore, tertiary phosphines are less π-acidic than carbon monoxide and formation of carbonyl-bridges is therefore to be expected from pure electronic effects. That substitution of carbonyl groups can take place without significant steric effects is shown by the structure of $FeCo_3(CO)_9[P(OMe)_3]_3H$ [132)], (Fig. 13).

This is a remarkable structure because, using high precision diffraction data obtained at −139 °C, it has been possible to localise directly the H atom, which makes the earlier claims for the hydrogen in $FeCO_3(CO)_{12}H$ being in the middle of the tetrahedron[189, 246)] seem less likely. Substitution of a carbonyl group under the basal plane containing the bridging carbonyl groups is also in agreement with the ^{13}C NMR spectrum of $Co_4(CO)_{11}[P(OMe)_3]$[66)], and the presence of bridging carbonyl groups has also been observed for the mono-substituted iridium compound

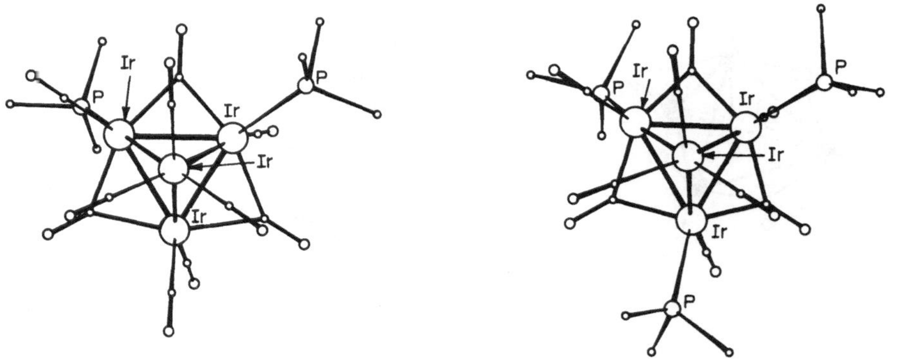

$\bar{d}_{Ir\text{-}Ir} = 2.73$ Å $\bar{d}_{Ir\text{-}CO_t} = 1.85$ Å $\bar{d}_{Ir\text{-}CO_b} = 2.1$ Å

Fig 14. Schematic structures of $Ir_4(CO)_7(\mu_2\text{-}CO)_3(PPh_3)_2$, (C_1) and $Ir_4(CO)_6(\mu_2\text{-}CO)_3(PPh_3)_3$, (C_s)

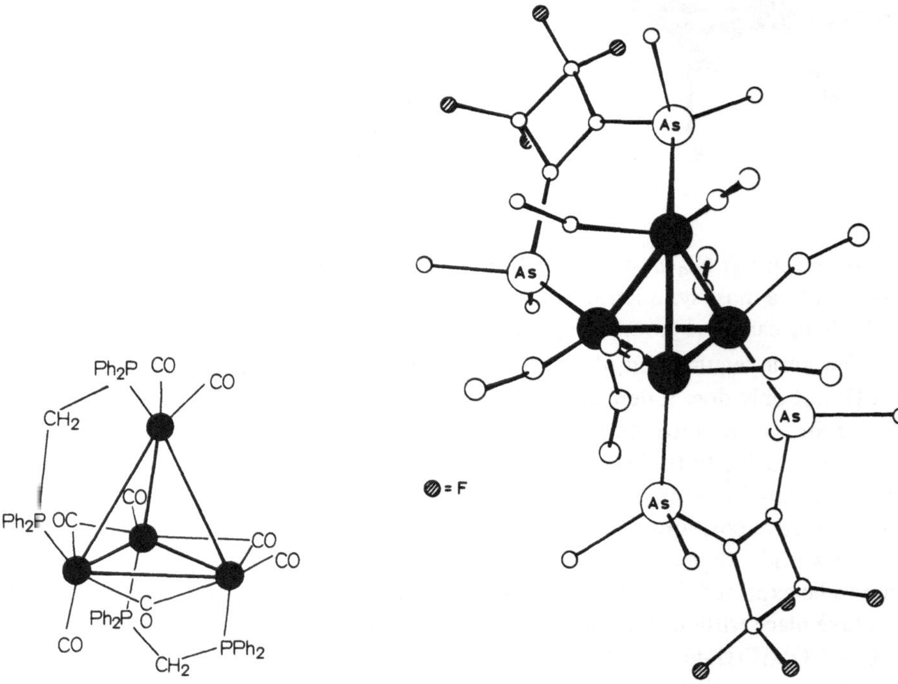

$\bar{d}_{Rh\text{-}Rh} = 2.711$ Å
$\bar{d}_{Rh\text{-}CO_t} = 1.91$ Å
$\bar{d}_{Rh\text{-}CO_b} = 2.09$ Å

$\bar{d}_{Co\text{-}Co} = 2.428$ Å

Fig. 15. Schematic structures of $Rh_4(CO)_5(\mu_2\text{-}CO)_3(DPM)_2$, (C_1) and $Co_4(CO)_8(ffars)_2$, (C_2)[43, 88]

$Ir_4(CO)_{11}(PPh_3)$[149]. The structures of the di- and tri-substituted iridium derivative are shown in Fig. 14. In all cases, the phosphines are around the basal plane. However, whereas the two phosphines are in relative *trans*-positions in the di-substituted compound, in the tris-derivative two of the phosphines are obliged to occupy relative *cis*-position and, as a result, are involved in more steric interaction.

Substitution of four carbonyl groups on the four different metal atoms is exemplified by the structure of $Rh_4(CO)_8(DPM)_2$[43] (Fig. 15) while the particular structure of $Co_4(CO)_8(ffars)_2$[88] (Fig. 15) is probably connected with the rigidity of the bidentate ligand, ffars.

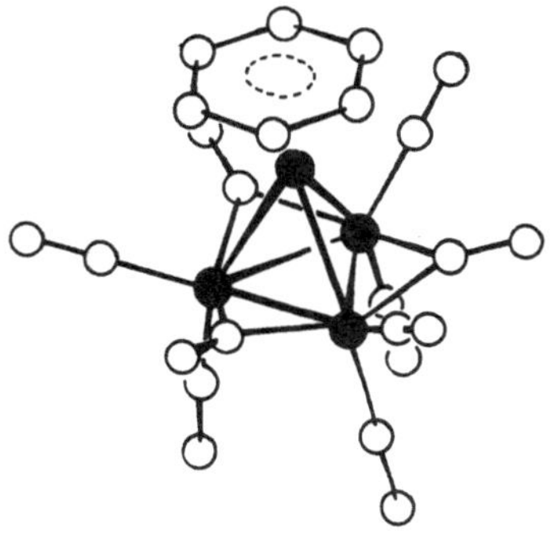

$\bar{d}_{Co-Co}$ = 2.481 (4) Å (apical), 2.455 (1) Å (basal)

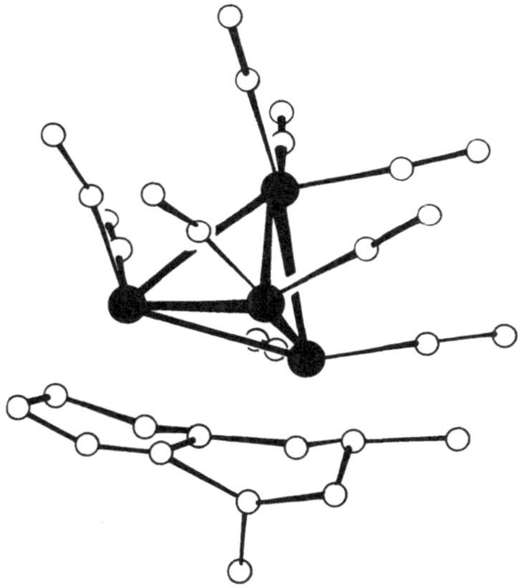

$\bar{d}_{Ru-Ru}$ = 2.752 Å (apical), 2.890 Å (basal)
$\bar{d}_{Ru-CO}$ = 1.861 Å

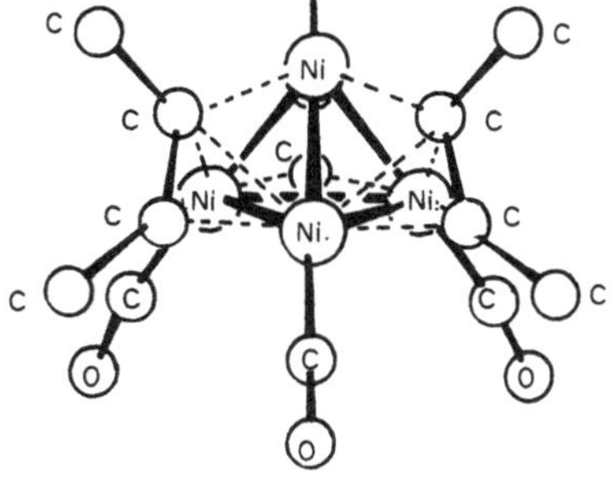

$\bar{d}_{Ni-Ni}$ = 2.52 Å
$\bar{d}_{Ni-CO}$ = 1.81 Å

Fig. 16. Schematic structures of $Co_4(CO)_6(\mu_2\text{-}CO)_3(mes)$, (C_{3v})[20], $Ru_4(CO)_9$(4,6,8-trimethyl-azulen) (C_{3v})[62, 63] and $Ni_4(CO)_4(\mu_3\text{-}CF_3C_2CF_3)_3$, (C_{3v})[79]

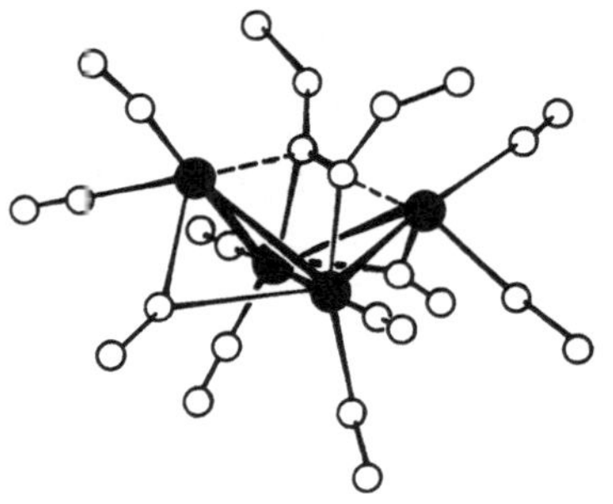

$\bar{d}_{Co-Co}$ = 2.552 Å (internal)
2.434 Å (external)

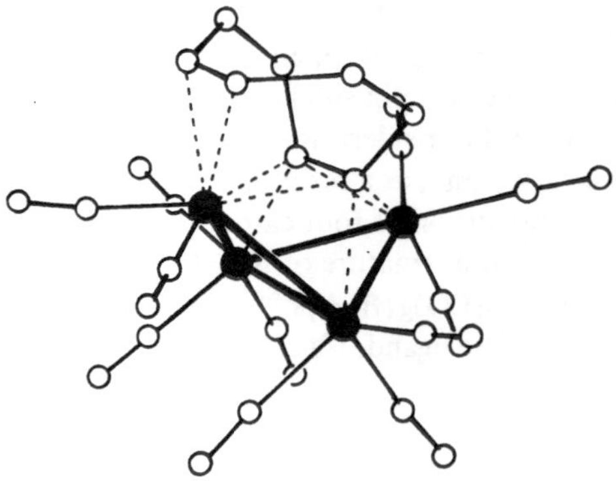

$\bar{d}_{Ru-Ru}$ = 2.823 Å (internal)
2.738 Å (external)

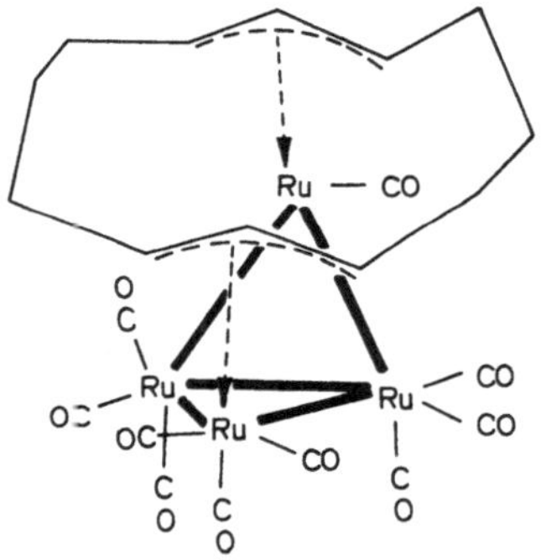

$\bar{d}_{Ru-Ru}$ = 2.850 Å (internal)
2.784 Å (external)

Fig. 17. Schematic representation of the "butterfly" structures found in $Co_4(CO)_8(\mu_2\text{-}CO)_2\text{-}(\mu_4\text{-}C_2Et_2)$, (C_2)[78], $Ru_4(CO)_{10}(\mu_4\text{-}C_{12}H_{16})$[16] and $Ru_4(CO)_{11}(\mu_4\text{-}C_8H_{10})$, (C_s)[186]

As is well known, all three carbonyl ligands on a $M(CO)_3$ group can be replaced by an arene as shown by the structure of $Co_4(CO)_9(mes)$[20], (Fig. 16) which is derived from $Co_4(CO)_{12}$. Figure 16 also shows that the more sterically demanding 4,6,8-trimethylazulene behaves as a tri-allyl ligand and is arched across a whole face of the tetrahedron in $Ru_4(CO)_9$(trimethylazulene)[62, 63]. The carbonyl groups also bonded to this face are compressed to C···C contacts of about 2.5 Å. This compound has been isolated as monoclinic and triclinic red crystals, but X-ray analysis on both types of crystals show that they contain indistinguishable molecular units.

The structure of $Ni_4(CO)_4(CF_3C_2CF_3)_3$ [79] in which each acetylenic ligand can be considered to provide 6 electrons to the cluster is also shown in Fig. 16.

Particular organic ligands such as acetylenes and allenes can induce opening of the tetrahedron to give a folded rhombus, which is usually classified as a "butterfly" structure (Fig. 17). This "butterfly" structure is also rather surprisingly found for the anion $[Fe_4(CO)_{13}H]^-$ [178], as shown in Fig. 18.

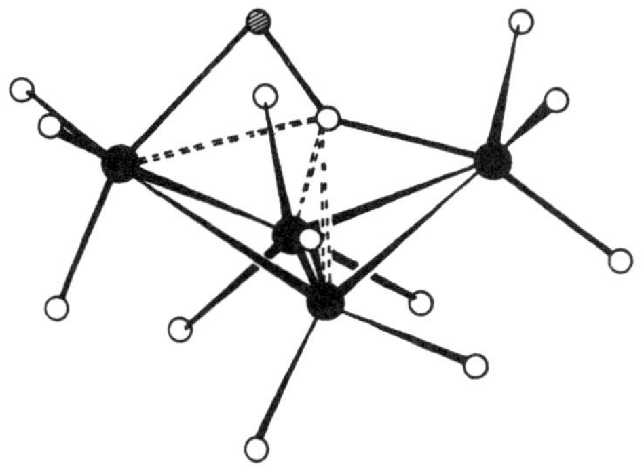

$\bar{d}_{Fe\text{-}Fe}$ = 2.63 Å
$\bar{d}_{Fe\text{-}CO_t}$ = 1.75 Å
● = oxygen atom

Fig. 18. Schematic structure of the anion $[Fe_4(CO)_{12}(\mu_4\text{-}CO)H]^-$, (C_s)[178)]

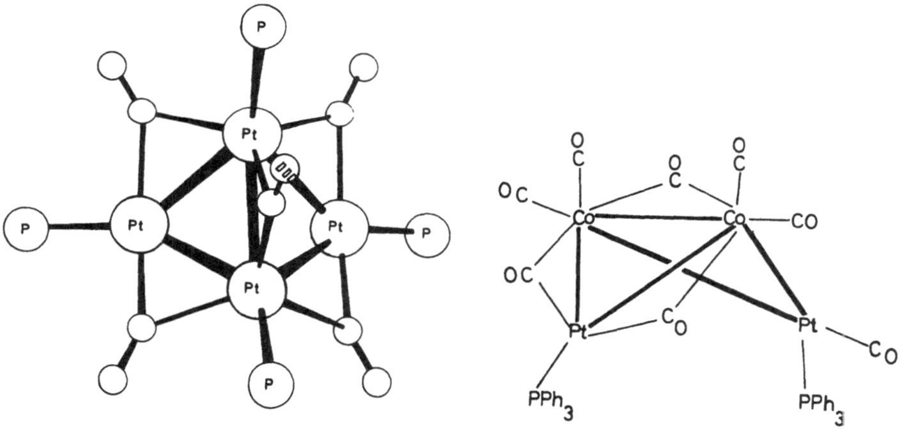

$\bar{d}_{Pt\text{-}Pt}$ = 2.790(7) Å (internal)
= 2.572 Å (external)

$\bar{d}_{Co\text{-}Co}$ = 2.498(3) Å $\bar{d}_{Pt\text{-}Co}$ = 2.550 Å

Fig. 19. Schematic structures of $Pt_4(\mu_2\text{-}CO)_5(PPhMe_2)_4$, (C_{2v})[239)] and $Pt_2Co_2(CO)_5(\mu_2\text{-}CO)_3\text{-}(PPh_3)_2$, (C_1)[95)]

It is worth noting that in this case the bridging carbon monoxide behaves as a 4-electron donor, which is similar to that previously observed in $Mn_2(CO)_4(\mu_2\text{-}CO)\text{-}(DPE)_2$ [67)].

Similar open structures containing platinum atoms with a 16 electron configuration have been found in $Pt_4(CO)_5(PPhMe_2)_4$ [239)] and in $Pt_2Co_2(CO)_8(PPh_3)_2$ [95)] as shown in Fig. 19.

A related structure has also been found for $Os_3Au(CO)_{10}(PPh_3)(\mu_2\text{-}X)$, (X = Cl or Br) [26)] and is shown in Fig. 20.

Further breaking of metal-metal bonds gives rise to the rectangular arrangement which is exemplified by the cobalt compounds, $Co_4(CO)_{10}(\mu_4\text{-}E)_2$ already shown in Fig. 1. It is interesting that on breaking another metal-metal bond, the three residual

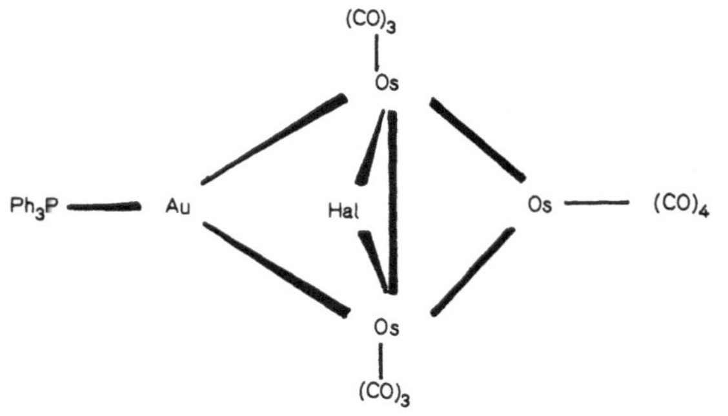

$\bar{d}_{Os\text{-}Os}$ = 2.85 Å

Fig. 20. Schematic structure of $Os_3Au(CO)_{10}(PPh_3)(\mu_2\text{-}X)$, (X=Cl, Br), (C_s) [26)]

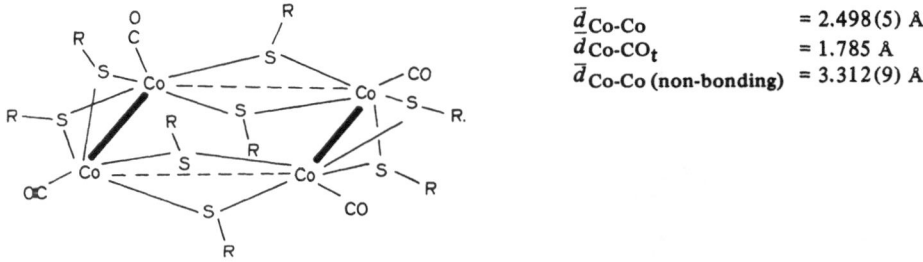

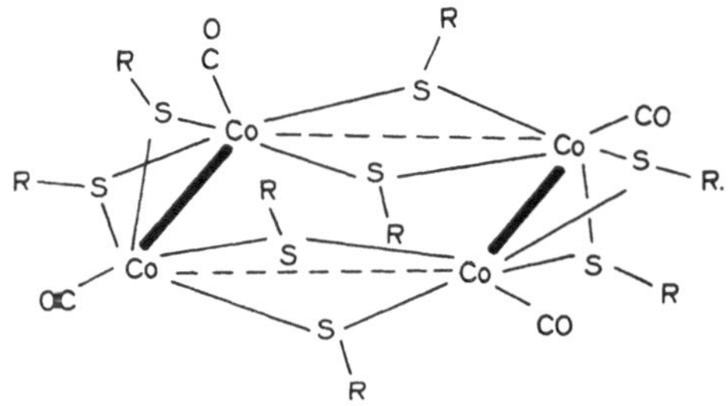

$\bar{d}_{Co\text{-}Co}$	= 2.498(5) Å
$\bar{d}_{Co\text{-}CO_t}$	= 1.785 Å
$\bar{d}_{Co\text{-}Co\ (non\text{-}bonding)}$	= 3.312(9) Å

Fig. 21. Schematic structure of $Co_4(CO)_4(\mu_2\text{-}SEt)_8$ [244)]

metal bonds connecting four metal atoms can be arranged to give an open tetrahedron as found in $Ni_4(t\,BuNC)_4(\mu_2\text{-}t\,BuNC)_3$, ($\bar{d}_{Ni\text{-}Ni_{bonding}}$ 2.48 Å, $\bar{d}_{Ni\text{-}Ni_{non\text{-}bonding}}$ 3.67 Å) [165)].

With only 2 metal-metal bonds, it is no longer possible to bond directly four metal atoms and we enter the border area between cage compounds and clusters. Examples of these compounds are $Co_4Cp_4(\mu_3\text{-}P)_4$ [222)] and $Co_4(CO)_4(\mu_2\text{-}SEt)_8$ [244)], the last structure being shown in Fig. 21.

B. Mössbauer

Mössbauer measurements have been made on a number of tetranuclear clusters (Table 7). However, the inequivalent iron atoms in $[Fe_4(CO)_{13}H]^-$ [93)], $CpRhFe_3(CO)_{11}$ [159)] and $Cp_2Rh_2Fe_2(CO)_8$ [159)] were not distinguished and the structural predictive capability of this technique is limited. Less ambiguous information can be obtained from the variation of the quadrupole splitting (Δ) within a closely related series of compounds since removal of pseudo-octahedral symmetry leads to larger quadrupole splittings. The change in Δ for the series of complexes $FeCo_3(CO)_{12-x}(PMePh_2)_xH$ ($x = 0\text{–}3$) is consistent with successive phosphine substitution of a carbonyl on each cobalt[70)], since small values of Δ are found when $x = 0$ or 3, which is consistent with a pseudo-octahedral symmetry, compared with

Table 7. ^{57}Fe Mössbauer data on tetranuclear clusters

Compound	T(°K)	δ (mm/sec)	Δ (mm/sec)	Ref.
$(Et_4N)_2[Fe_4(CO)_{13}]$	80	0.28[1]	0.27	93)
(pyH) $[Fe_4(CO)_{13}H]$	80	~0.33[1]	~0.67	93)
$Fe_4(CO)_4Cp_4$	80	0.66[1]	1.76	103)
$[Fe_4(CO)_4Cp_4]Cl$	80	0.67[1]	1.38	103)
$[Fe_4(CO)_4Cp_4]Br_3$	80	0.67[1]	1.40	103)
$[Fe_4(CO)_4Cp_4]I_5$	80	0.67[1]	1.42	103)
$CpRhFe_3(CO)_{11}$	293	0.22[1,4]	0.89	159)
$Cp_2Rh_2Fe_2(CO)_8$	293	0.27[1,4]	0.80	159)
$(Et_4N)[FeCo_3(CO)_{12}]$	295	0.05[2]	0.14	70)
$FeCo_3(CO)_{12}H$	295	0.12[2]	0.08	70)
$FeCo_3(CO)_{11}(PMePh_2)H$	295	0.10[2]	0.36	70)
$FeCo_3(CO)_{10}(PMePh_2)_2H$	295	0.12[2]	0.35	70)
$FeCo_3(CO)_9(PMePh_2)_3H$	295	0.11[2]	0.10	70)
$FeCo_3(CO)_9(DPE)_2H$	295	0.11[2]	0.11	70)
$FeCo_3(CO)_8(DPE)_2H$ (B)[3]	295	0.13[2]	0.39	70)
$FeCo_3(CO)_8(DPE)_2H$ (A)[3]	295	0.08[2]	1.01	70)

1) Relative to sodium nitroprusside.
2) Relative to iron foil.
3) (A) and (B) are different isomers.
4) Relatively large half-widths.

the large values obtained when $x = 1$ or 2; the pseudo-octahedral symmetry, when $x = 3$, could also be produced by substituting all 3 carbonyls on iron by the tertiary phosphine, but this is discounted on steric and electronic ground.

C. Raman

The Raman spectra of solid samples give useful information concerning the mode of bonding of the hydride (or deuteride) group in a cluster; bands associated with terminal hydride- and bridging hydride-metal bonds occur at $1900 \pm 300\ cm^{-1}$ and $1100 \pm 300\ cm^{-1}$ respectively and are generally stronger in the Raman than in the infrared spectra[147]. Metal-metal bonds exhibit intense bands in Raman spectra [for instance recent work reports absorptions at 250 and 185 cm^{-1} for $Co_4(CO)_{12}$ in acetone[205a]] and, in conjunction with infrared studies, metal-metal force constants for $Ir_4(CO)_{12}$ (1.69 mdyn/Å)[214] and $[Fe(CO)Cp]_4$ (1.3 mdyn/Å)[230] have been evaluated, although the value of f(Ir-Ir) obtained for $Ir_4(CO)_{12}$ seems disproportionately large when compared with f(Os-Os) in $Os_3(CO)_{12}$ (0.91 mdyn/Å)[213]. This difference is probably a reflection of the greater complexity of the force field in $Ir_4(CO)_{12}$ and illustrates the problems involved in evaluating interaction constants in these computations as the size of the cluster increases. The high intensity of the bands due to metal-metal vibrations is generally because of irradiation within or near an electronic transition involving the metal framework, which leads to enhanced Raman scattering *via* resonance mechanisms. However, an unwanted aspect of this electronic absorption is sample decomposition, which may be alleviated by sample spinning[152].

VI. Structural Data in Solution

A. Infrared

Because of the extreme experimental simplicity and the high sensitivity in the carbonyl region, infrared spectroscopy is most important in the study of these compounds. The progressive shift towards lower frequencies, which is observed on increasing the negative charge is illustrated by the data in Table 8, while Table 9 illustrates the similar effect which is observed on progressive substitution with less π-acidic ligands.

Infrared spectroscopy has often been used for obtaining structural information but, as shown in the long debate about the structure of $Co_4(CO)_{12}$ in solution[66, 92], only partial information is generally obtained and unequivocal interpretations are rarely possible.

The application of vibrational spectra to metal carbonyls has already been discussed[19, 29, 105] and rigorous assignments of the carbonyl bands in the infrared spectra of tetranuclear clusters are difficult to make because of many possible alternatives. This is accentuated by the practical difficulties in obtaining Raman spectra of solutions of these clusters. The infrared spectra of samples with various isotopic enrichments should, in principle, assist with these assignments but little work in this area has yet been reported. The most rigorous assignments of carbonyl bands have been made for $Co_3(CO)_9$ arene (arene = toluene, mesitylene, tetralin)[23], $FeCo_3(CO)_{12}H$[25] and $M_4(CO)_{12}$ (M = Co, Rh)[25]. The force and interaction constant calculations are usually simplified by the fact that the interaction constants between bridging and terminal carbonyls appear to be zero. However, all other interactions have to be considered, with the geminal interaction constants (0.25 – 0.31 mdyn/Å) being the biggest. There is no general trend in the values of the force constants obtained for the axial, equatorial and apical terminal carbonyls, which are in the range 16.34–17.13 mdyn/Å. As expected, these values are considerably higher than the force constants for bridging carbonyls, [$Co_4(CO)_{12}$ 14.23 mdyn/Å and $FeCo_3(CO)_{12}H$, 14.55 mdyn/Å[25]].

B. Nuclear Magnetic Resonance

^1H NMR has long been routinely used to detect the presence of M-H bonds and also to obtain structural information about clusters containing organic ligands. The availability of Fourier transform NMR spectrometers over the last few years has enabled studies involving ^{13}C, ^{31}P and other nuclei also to be carried out. This has made NMR an extremely useful technique for obtaining structural information and, be-

Table 8. Examples of the progressive shift of the carbonyl stretching frequency on increasing the negative charge on the cluster

Compound	Solvent	ν(CO) cm^{-1}
$Rh_4(CO)_{12}$	Heptane[42)]	2101 (w), *2074.5 (vs)*, *2069 (vs)*, 2059, 2044.5 (m), 2041.5 (m), 2001 (w), 1918.4 (w), 1882 (s), 1848 (w)
$[Rh_4(CO)_{11}(COOCH_3)]^-$	THF[185)]	2075 (w), *2030 (vs)*, *2010 (s)*, 1995 (sh), 1965 (sh), 1892 (w), 1845 (s)
$[Rh_4(CO)_{11}]^{2-}$	THF[185)]	1984 (vw), *1930 (vs)*, *1810 (s)*
$Re_4(CO)_{12}H_4$	Cyclohexane[218)]	*2090 (s)*, *1985 (s)*
$[Re_4(CO)_{12}H_6]^{2-}$	Acetone[145)]	*2000 (s)*, *1910 (vs)*
$[Re_4(CO)_{15}H_4]^{2-}$	Acetone[5)]	2050 (w), *2000 (vs)*, *1960 (s)*, 1920 (m), 1890 (m), 1880 (sh)

Table 9. Comparison of the frequencies corresponding to the stronger infrared absorptions (cm^{-1}) for the terminal carbonyls in $Co_4(CO)_{12-n}L_n$

Compound	Unsubstituted[168)]	PEt_3[70, 168)]	PMe_2Ph[168)]	$P(OMe)_3$[168)]	*t* BuNC[202)]
$Co_4(CO)_{12}$	2062.5 2053.7 (a)	–	–	–	–
$Co_4(CO)_{11}L$	–	2038 2027 (b)	2042 2036 (a)	2047.5 2042 (a)	2046 2040 (b)
$Co_4(CO)_{10}L_2$	–	2020 2016 (a)	2021 2014 (a)	2035 2016 (a)	2027 2013 (b)
$Co_4(CO)_9L_3$	–	1978 (a)	1995 1979 (a)	2010 2003 (a)	2011 2002 (b)
$Co_4(CO)_8L_4$	–	–	–	1980 1950 (a)	1997 1976 (b)

(a): CH_2Cl_2 (b): hexane

Table 10. High field ^1H NMR data for tetranuclear hydride clusters. Figures in parentheses refer to relative intensities and d = doublet, t = triplet, qu = quintet

	Colour	Spectrum recorded at T °C	τ_H		Ref.
$Re_4(CO)_{16}(OMe)H_2$	Cream	38	17.2		5)
$[Re_4(CO)_{15}H_4]^{2-}$	Yellow	38	20.5 (2), 26.9 (2)		5)
		−39	15.0 (1), 25.9 (1), 26.9 (2)		
$Re_4(CO)_{12}H_4$	Deep red	R.T.	15.1		147)
$[Re_4(CO)_{12}H_6]^{2-}$	Light yellow	R.T.	27.4		145)
$FeRu_3(CO)_{13}H_2$	Orange	R.T. to −120	28.5		163, 252)
$FeRu_3(CO)_{12}H_4$	Dark red	R.T. to −130	29.3		163)
$[Fe_3Ni(CO)_{12}H]^-$		R.T.	30.5		171)
α-$Ru_4(CO)_{13}H_2$	Red	R.T.	28.5, 28.6		140, 163)
β-$Ru_4(CO)_{13}H_2$	Red	R.T.	19.1		138)
$[Ru_4(CO)_{12}H_3]^-$	Orange/red	R.T.	26.9		164)
		−95	25.9 (2) d, 27.4 (4), 29.0 (1) d,	J(H-H) 2.5 Hz	
α-$Ru_4(CO)_{12}H_4$	Yellow	R.T.	27.6 – 28.0		140, 138, 163, 210)
"β-$Ru_4(CO)_{12}H_4$"	Yellow	R.T.	18.6		138)
$[Ru_4(CO)_{12}(PhC_2Ph)H]^+$	Brown	+10 to −60	33.4 – 33.9		142)
$[Ru_4(CO)_{12}(PhC_2Me)H]^+$	Brown	0 to −60	32.8		142)
$Ru_4(CO)_{11}(PBu_3)H_4$	Orange/red	R.T.	27.7		210)
$Ru_4(CO)_{11}(PPh_3)H_4$	Yellow/brown	R.T.	27.7		210)
$Ru_4(CO)_{11}[P(OMe)_3]H_4$	Orange	R.T.	27.7 d	J(P-H) 2.6 Hz	162)
		−124	27.5	J(P-H) 2.5 Hz	
			28.1	J(P-H) 12.5 Hz	148)
$Ru_4(CO)_{10}(PPh_3)_2H_4$	Brown/red	R.T.	26.1, 26.9		210)
$Ru_4(CO)_{10}[P(OMe)_3]_2H_4$	Orange	R.T.	27.6 t	J(P-H) 6.63 Hz	162)
$Ru_4(CO)_9(PBu_3)_3H_4$	Deep red	R.T.	26.5		210)
$Ru_4(CO)_9(PPh_3)_3H_4$	Deep red	R.T.	26.1, 27.1		210)
$Ru_4(CO)_9[P(OMe)_3]_3H_4$	Orange	R.T.	27.8 q	J(P-H) 7.70 Hz	162)

$Ru_4(CO)_9[P(OPh)_3]_3H_4$	Orange	R.T.	27.3		36)
$Ru_4(CO)_8(PBu_3)_4H_4$	Orange	R.T.	26.8		210)
$Ru_4(CO)_8(PPh_3)_4H_4$	Red	R.T.	25.6		210)
$Ru_4(CO)_8[P(OMe)_3]_4H_4$	Orange	R.T. to −100	27.8 qu	J(P-H) 7.95 Hz	162)
$Os_2Pt_2(CO)_8(PPh_3)_2H_2$	Yellow/orange	R.T.	17.8	J(Pt-H) 524 Hz	34)
$Os_4(CO)_{13}H_2$		R.T.	31.2		87)
$Os_4(CO)_{12}H_4$	Colourless	R.T.	30.3		139)
$[Ir_4(CO)_{12}H_2]^{2+}$		R.T.	30.0		160)
$Ir_4(CO)_{11}H_2$		R.T.	25.5		9)
$[Ir_4(CO)_8(AsMePh_2)_4H]^+$		R.T.	30.8		46)
$[Ir_4(CO)_8(AsMePh_2)_4H_2]^{2+}$		R.T.	27.5		46)
$[Ir_4(CO)_8(PMe_2Ph)_4H_2]^{2+}$		R.T.	27.5	J(P-H) 5.6 Hz	46)

cause stereochemical non-rigidity, which is of interest in itself, is often found in this area, the study of ligand exchange processes.

^{1}H NMR

Metal-hydrogen bonds generally give a high field signal in the ^{1}H NMR spectrum in the region 15–35 τ as illustrated by the data on tetranuclear clusters, which is summarised in Table 10. The presence of quadrupolar nuclei $\left(e.g.\ ^{59}\text{Co}\ I = \frac{7}{2}\right)$ can, however, lead to unfavourable relaxation effects as exemplified by the non-observance of a high field signal for $Fe_3Co(CO)_{12}H$[189].

Terminal hydrides generally occur at lower fields than hydrides occupying bridging positions, as discussed in Kaesz and Saillant's comprehensive review on metal hydrides[147] and as shown by the self-consistent X-ray and ^{1}H NMR data recently obtained for $[Re_4(CO)_{15}H_4]^{2-}$ (terminal τ_H 15.04, edge-bridging τ_H 25.95, 26.93)[5] and $Os_3(CO)_{11}H_2$ (terminal τ_H ca. 20, edge-bridging τ_H ca. 30)[220]. However, this rule must be used with caution because of the possibility of intra-hydrogen exchange. In principle this should be detected by the dependence of the spectra on temperature but actually when the number of high field resonances is less than the number of hydrogens present in multihydrido clusters, it is often difficult to ascertain whether this is because the hydrogens are:

a) fluxional,

b) occupying inequivalent sites, which give rise to accidentally coincident NMR signals, or

c) occupying equivalent sites.

The latter reason is clearly in keeping with the single resonance observed for $FeRu_3(CO)_{13}H_2$ at $-120°$[163] and the X-ray structure (Fig. 9)[100], but it is difficult to decide whether a) or b) is responsible for $FeRu_3(CO)_{12}H_4$ giving a single resonance even at -130 °C.

Introduction of a different magnetic nucleus into the cluster can provide additional information about hydrogen exchange. At room temperature, rapid intramolecular hydrogen exchange is found in the phosphite substituted derivatives $Ru_4(CO)_{12-x}[P(OMe)_3]_xH_4$ ($x = 1,2,3,4$)[162] and $Ru_4(CO)_{13-y}[P(OMe)_3]_yH_2$ ($y = 1,2$)[148] since each gives a single high field signal with an averaged coupling to the phosphorus nuclei whereas, on going to low temperature (-124 °C), the spectrum of $Ru_4(CO)_{11}[P(OMe)_3]H_4$ (Table 10) becomes consistent with the predominance of just one isomer [Fig. 22, $L_1 = P(OMe)_3$, $L_{2,3,4} = CO$][163]. This should be contrasted with $[Ru_4(CO)_{12}H_3]^-$ which is found to exist as two isomers at low temperatures (-95 °C); one is of C_{3v} symmetry and the other of either C_2 or C_{2v} symmetry[164]. Both forms are rapidly interconverting and their relative concentrations vary with solvent. The activation energy for intrahydrogen exchange in $Ru_4(CO)_{11}[P(OMe)_3]H_4$ is 3.4 kcal mol^{-1} [163] which is similar to that found for the diffusion of hydrogen in metals[226].

The above studies are important in providing a start towards an understanding of energy barriers involved in hydrogen rearrangements in tetranuclear hydrido clusters and make the earlier claims for the existence of $Ru_4(CO)_{13}H_2$ and $Ru_4(CO)_{12}H_4$ as α- and β-isomers at room temperature[138] look rather doubtful, especially as recent work on $Ru_4(CO)_{12}H_4$ failed to find any evidence for more than one isomer[163, 210]. However, much more work is clearly required before we have a better understanding of the dependence of hydrogen exchange on coordination numbers, steric and electronic factors.

^{13}C NMR

The number of ^{13}C NMR studies on metal carbonyls has greately increased over the last few years as shown by the presently available data on tetranuclear carbonyls (Table 11) compared with that available in 1974 when the last review[232] on this subject appeared and contained only two references to tetranuclear clusters.

Before discussing the results in Table 11 it is worthwhile mentioning a few points of general importance. Measurements are usually made on solutions containing a relaxing agent [*e.g.* $Cr(acac)_3$ [d)]][97] together with the ^{13}CO-enriched sample, which can often be prepared by direct exchange with ^{13}CO gas. The ratio of the concentration of relaxing agent to carbonyl should be kept as low as possible (1 : 100 mol ratio)[75] in order to avoid line-broadening and, in the absence of exchange processes, reasonably sharp signals are usually obtained because $^2J(^{13}C\text{-}^{13}C) < 10$ Hz. Increased back-bonding results in a shift of δ_{CO} to lower field[110] as exemplified by the progressive shift with increasing bridging character ($\delta_{\mu_1\text{-}CO} < \delta_{\mu_2\text{-}CO} < \delta_{\mu_3\text{-}CO}$) and, within a triad, there is a progressive shift to higher field with increasing atomic number[232].

Line-broadening, due to quadrupolar relaxation, often prevents the observance of room temperature ^{13}C NMR spectra of cobalt carbonyls but it is significantly reduced at lower temperatures. Nevertheless, the lines are still broad and it would be highly desirable to record the ^{13}C NMR spectra of cobalt carbonyls with ^{59}Co-spin-decoupling, which might also help to explain the anomalous peak intensities that are sometimes found. Thus, whereas the peak intensities in the low temperature spectra of $Co_4(CO)_9(C_6H_5Me)$ [92], $Co_3Rh(CO)_{12}$ [141] and $Co_4(CO)_{11}[P(OMe)_3]$ [66] are as expected, $Co_4(CO)_{12}$ [66, 92] gives *three* equally intense peaks instead of the *four* equally intense peaks which are expected as a result of ^{59}Co NMR studies[72] which suggested that the solid state structure[243] (Fig. 4) is retained in solution.

The low temperature ^{13}C NMR spectrum of $Rh_4(CO)_{12}$, which is *iso*-structural with $Co_4(CO)_{12}$ in the solid state, has three equally intense doublets due to the apical, axial, equatorial carbonyls and a triplet due to bridging carbonyls[91]. These multiplets arise through spin-spin coupling to ^{103}Rh (100% abundant, $I = \frac{1}{2}$), and $^1J(Rh\text{-}CO)$ is larger for terminal than bridging carbonyls because of a higher *s*-character in the Rh-C bond (see Table 11). On increasing the temperature all the peaks broaden at the same rate due to intra-carbonyl exchange until at 50–60 °C a quintet is observed[73] as a result of *fast* exchange of CO from one site to another

[d)] Tris(acetylacetonato)-chromium(III).

Table 11. ^{13}C NMR data for tetranuclear carbonyl clusters. δ_{CO} (p.p.m.) with J(M-CO) in parentheses (Hz) and relative intensities in [ ]

Compound	T°	μ_3-CO	μ_2-CO	μ_1-CO	Ref.
$[Fe_4(CO)_{13}]^{2-}$	R.T.		229.75		111)
$FeRu_3(CO)_{12}H_2$	−92		231.9 [2]	212.0, [1] 204.4, [1] 195.3, [2] 191.1, [2] 190.4, [2] 188.2, [1] 187.7 [2]	193)
	+95			199.5	
$Ru_4(CO)_{12}H_4$				188.7	193, 2)
$[FeCp(CO)]_4$	R.T.	289.8			232)
$Co_4(CO)_{12}$	−95 to −30		244 [3]	196 [3] 192 [3]	66)
	−100 to −60		243.1 [3]	195.9 [3] 191.9 [3]	92)
$Co_3Rh(CO)_{12}$	−85		251.2, [1] 238.3 (38) [2]	201.1, [2] 195.5, [3] 188.2(78), [1] 183.1 (51) [1]	141)
	+30			201.3	
$Rh_4(CO)_{12}$	−65		228.8 (35) [3]	183.4(75), [3] 181.8(64), [3] 175.5 (62) [3]	91)
	+50			189.5 (17.1)	
$[Rh_4(CO)_{11}(COOMe)]^-$	−72 to −30			201.7 (18.6)	111)
$[Rh_4(CO_{11}]^{2-}$	−72			222.8 (19.5)	111)
$Co_4(CO)_{11}[P(OMe)_3]$	−82		248.2 [2] 245.6 [1]	198.0 197.2 [3] 196.1 [3] 193.7 [2]	66)

$Co_4(CO)_{10}[C_2(COOMe)_2]$	−104			211.2 [4]	198.2 [2]	191.3 [4]	92)
	−20			203			
$Co_4(CO)_{10}(C_2Ph_2)$	−90			213.6 [4]	198.7 [2]	193.5 [4]	46)
$Co_4(CO)_9(C_6H_5Me)$	−70	247.3 [3]		198.7 [3]	180.9 [3]		92)
$Ir_4(CO)_8(AsMe_2Ph)_4$	R.T.	221.2 [2]		176.6 [2]	169.6 [2]	137.2 [2]	92)
	+100			176.1			
$Ir_4(CO)_8(AsMePh_2)_4$	R.T.	223.7, [1]	220.3 [2]	175.5, [2]	169.1, [1]	135.6 [2]	46)

and from one rhodium to another. The observed chemical shift of the quintet is in agreement with the calculated average and, since $^2J(Rh\text{-}CO_t)$ is small, the observed spacing in the quintet is found to be very close to $^1J(Rh\text{-}CO_t)/n$, where n = number of rhodium atoms (in this case 4).

The barrier to carbonyl scrambling is reduced considerably on increasing the negative charge on the cluster as found for $[Rh_4(CO)_{11}(COOMe)]^-$ [111] and $[Rh_4(CO)_{11}]^{2-}$ [111] which, even at –70 °C, give quintets (Table 11) due to rapid carbonyl intra-exchange; the carbomethoxy group in $[Rh_4(CO)_{11}(COOMe)]^-$ is not involved in this scrambling process and gives rise to a doublet [δ 181.47 p.p.m.; ^1J(Rh-COOMe) 40.0 Hz][111]. Although this increased fluxionality can in part be attributed to the increased nucleophilicity of the metal skeleton which can accumulate on particular metals with concurrent formation of carbonyl bridged intermediates, it is difficult to assess how important this is, since changes in the geometrical distribution of carbonyl groups also result on increasing the charge (for instance compare Figs. 4 and 11). Nevertheless, this probably accounts for the selective carbonyl scrambling in $Co_3Rh(CO)_{12}$ [141] and $FeRu_3(CO)_{13}H_2$ [193]. In both cases, the low temperature limiting spectra are consistent with the instantaneous structures and carbonyl scrambling is found to occur about the first row elements, before the second row elements, and lastly over the whole tetrahedron.

Attempts to obtain stereochemical information on $Ir_4(CO)_8L_4$ (L = tertiary phosphine or arsine) from NMR studies did not give unambiguous results and the carbonyl chemical shifts at 135–137 p.p.m. [46] seem anomalously high.

^{59}Co and ^{31}P NMR have not yet been much used for the characterization of tetranuclear clusters but, in conjunction with NMR studies involving other nuclei, appear to offer great potential.

The only ^{59}Co NMR study is on $Co_4(CO)_{12}$ and the apparent discrepancy between the results obtained by two independent groups in 1967 [104, 173] has recently been resolved[66]. The ^{59}Co NMR spectrum of a solution of $Co_4(CO)_{12}$ in hexane at 30 °C shows two peaks at 8,400 and 9,670 p.p.m. to high field of $[Co(NH_3)_6]Cl_3$ with relative intensities 1 : 3 respectively[66]. This is clearly consistent with the solid state structure (Fig. 4) but is inconsistent with the ^{13}C NMR results[66, 92].

^{31}P NMR data have been reported for $Ir_4(CO)_8(PPh_2Me)_4$ [δ –8.4(1), –40.6(2), 61.6(1) p.p.m.] together with a variable temperature ^{1}H NMR study, which indicates that phosphine intra-exchange occurs at high temperatures whereas the low temperature spectra are consistent with a number of alternative stereochemical arrangements[46].

C. Electron Spin Resonance

The preparation and ESR measurements of two tetranuclear carbonyl cluster radical anions have been reported: $[Ir_4(CO)_{12}]^-$, (g = 2.002) [208] and $[Fe_4(CO)_4Cp_4]^-$, (g = 2.013) [94] while it has not been possible to observe a signal for the cation $[Fe_4(CO)_4Cp_4]^+$ [94].

VII. Mass Spectrometry

Several reviews on mass spectrometry of organometallic compounds are available[143, 152, 170], and therefore we will only point out some general observations. Mass spectrometry has the advantage of only requiring small amounts of compound and is widely used for the determination of molecular formulae of uncharged carbonyl clusters, although at high molecular weights an internal standard is generally required. However, this technique must be used with caution since misleading results can be obtained due to the ready loss of a ligand, which results in the parent molecular ion *not* being observed. This is vividly illustrated by the X-ray structure of the tetracarbonyl cluster, $Ni_4(CO)_4(CF_3C_2CF_3)_3$ [79] which, on the basis of mass spectra, was originally suggested to be a tricarbonyl $Ni_4(CO)_3(CF_3C_2CF_3)_3$ [153]. Similar problems can also arise in multihydrido carbonyl clusters[163].

In mixed metal clusters, redistributions can occur in the mass spectrometer so that the highest observed peak arises from such a product. This is clearly exemplified by the heterometallic dodecacarbonyls containing metals of the cobalt triad. The thermal stability of this series of clusters decreases as the rhodium content increases and facile redistributions result in the formation of tetranuclear species containing less rhodium; the very thermally stable $Co_2Ir_2(CO)_{12}$ does not rearrange[184].

VIII. Reactivity

Although it is true that there are some general reactions and trends in reactivities, it is necessary to emphasize the limits of such generalisations. Already a superficial analysis shows that there are large differences in the chemical behaviour of analogous compounds such as $Co_4(CO)_{12}$ and $Rh_4(CO)_{12}$ and strictly speaking the chemistry of each compound should be discussed *per se.* Furthermore, our knowledge of the mechanism of reaction of tetranuclear carbonyl clusters is very limited. Finally the high delocalization, which is generally believed to be present in these clusters, implies a high polarizability and makes most of the possible relationships between reactivity and physical properties of the clusters ambiguous.

The experimental evidence shows that the carbon of terminal carbonyl groups is positively polarized (or polarizable) and, contrary to the behaviour of free carbon monoxide, is easily attacked by strong nucleophiles (OH^-, OR^-); a behaviour which is general in the chemistry of metal carbonyls. Moreover, the negative polarisation (or polarizability) of oxygen atoms of carbonyl groups, particularly bridging carbonyls, is illustrated by the facile formation of adducts with Lewis acids as shown in *Eq.* (16) [7].

$$Fe_4(\mu_3\text{-}CO)_4 + 4\ AlEt_3 \xrightarrow[\text{toluene}]{25\ °C} Fe_4(\mu_3\text{-}CO\text{-}AlEt_3)_4Cp_4 \qquad (16)$$

$$[\nu(CO) = 1623\ cm^{-1}] \qquad\qquad [\nu(CO) = 1547\ cm^{-1}]$$

There are only a few examples of cationic clusters, which is in keeping with the postulated electron-sink effect of the system of metal-metal bonds. This hypothesis is supported by examination of some properties of the clusters $Co_3(CO)_9(\mu_3\text{-}CR)$ as reported previously[207, 209].

Increasing the negative charge on the cluster (*e.g.* on going from $FeCo_3(CO)_{12}H$ to $[FeCo_3(CO)_{12}]^-$) results in slower substitution reactions and this effect is believed to be due to the increase in back-donation with associated increase in M-CO bond energy[69].

For the compounds $Ir_4(CO)_{12-n}L_n$, it has been proposed that formation of bridges produces a further contribution to the polarizability of the cluster and makes successive substitution reactions easier[149]:

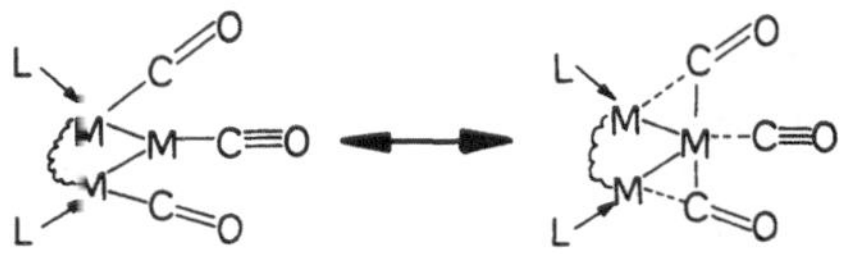

(17)

The reactivity is additionally complicated by the stereochemical nonrigidity of carbon monoxide, and also by the metallic bonds which, in agreement with their low energy, can open and close with surprising ease as is shown in the following example[178]:

$$[Fe_4(CO)_{13}]^{2-} \underset{t\,BuO^-}{\overset{H^+}{\rightleftharpoons}} [Fe_4(CO)_{13}H]^- \quad (18)$$

(tetrahedron, 6 Fe-Fe bonds) ("butterfly", 5 Fe-Fe bonds)

A. Reduction

Electrochemical studies[192] show that reduction of polynuclear systems involve intermediate formation of radical anions, *e.g.*[94].

$$Fe_4(CO)_4Cp_4 \xrightarrow[+e]{-1.3\ V} [Fe_4(CO)_4Cp_4]^- \qquad [\nu(CO)\ 1576\ cm^{-1}] \quad (19)$$

The observation of radical anions has been confirmed by ESR measurements as illustrated by $[Ir_4(CO)_{12}]^-$, ($g = 2.002$)[208]. Similarly, a toluene solution of $Co_4(CO)_{12}$ reacts with cobaltocene precipitating a brown compound which is extremely reactive and contains a cobaltocenium cation for each four cobalt atoms of the anion[55]. With excess cobaltocene (or alkali metals) in THF the reaction proceeds further as shown in Eq. (20),

$$Co_4(CO)_{12} \longrightarrow [Co_4(CO)_{12}]^-\ (?) \longrightarrow [Co_4(CO)_{11}]^{2-}\ (?) \longrightarrow [Co_6(CO)_{15}]^{2-} \longrightarrow [Co_6(CO)_{14}]^{4-} \quad (20)$$

and it seems probable that formation of the hexanuclear cluster is due to reaction of the intermediate tetranuclear dianion with the available carbon monoxide as found for $[Rh_4(CO)_{11}]^{2-}$, [Eq. (5)].

As already shown in Eqs. (3), (4) and (5), fragmentation of carbonyl cluster anions, on reaction with carbon monoxide becomes easier with increasing negative charge. It is therefore not surprising that reduction of tetranuclear clusters with alkali metals in excess, gives rise to simple mononuclear derivatives $[M(CO)_4]^-$, as found for $Co_4(CO)_{12}$ in liquid NH_3 [13, e)], $Rh_4(CO)_{12}$ in THF under CO[56)] and $Ir_4(CO)_{12}$ in THF under CO[175)].

A different method of reduction involves the reaction with strong base, such as OH^- and OR^-. In this case, the primary reaction is nucleophilic attack on a carbon atom of a terminal carbonyl group as shown by the formation of a carboalkoxy group on reaction with alkoxide:

$$M_4(CO)_{12} + OR^- \xrightarrow[ROH]{25\,°C} [M_4(CO)_{11}(COOR)]^- \qquad (M = Co, Rh, Ir) \qquad (21)$$

In the case of cobalt (R = *i*-Pr)[61)], the anion has a low stability and decomposes in a few hours, probably by demolition with carbon monoxide. With rhodium[185)] and iridium[102)] (R = OMe^-), it has been possible to isolate the carbomethoxy anion in a pure form. The carboalkoxy group undergoes further reaction with sodium hydroxide to give the corresponding dianion[185)]:

$$[Rh_4(CO)_{11}(COOMe)]^- + 3\,OH^- \longrightarrow [Rh_4(CO)_{11}]^{2-} + MeOH + CO_3{}^{2-} + H_2O \qquad (22)$$

Another useful reducing agent is sodium tetrahydroborate, $NaBH_4$, although in this case reduction can be complicated by progressive substitution of carbonyl groups with hydrogen atoms[147)].

B. Oxidation

Electrochemistry shows that oxidation also takes place in successive steps[94)].

$$Fe_4(CO)_4Cp_4 \xrightarrow[-e]{0.32\,V} [Fe_4(CO)_4Cp_4]^+ \xrightarrow[-e]{1.08\,V} [Fe_4(CO)_4Cp_4]^{2+} \qquad (23)$$

and it is worth noting that the dicationic species has a remarkably low stability and easily disproportionates.

An indirect method of oxidation consists of the intermediate formation of hydrido derivatives followed by hydrogen elimination. An example of this type of behaviour is shown by the following reactions[9)]:

---

e) In this particular case some carbon monoxide can be provided by the disproportionation of $Co_4(CO)_{12}$ [Eq. (52)].

$$[Ir_4(CO)_{11}H]^- + H^+ \xrightarrow{CO} Ir_4(CO)_{12}$$

$$[Ir_4(CO)_{11}H]^- + H^+ \xrightarrow{N_2} Ir_4(CO)_{11}H_2 \text{ (52\% yield)} \qquad (24)$$

A similar elimination of methanol is possibly involved in the transformation of the not completely characterized species, $[Re_4(CO)_{16}(OMe)H_4]^{3-}$, into $[Re_4(CO)_{15}H_4]^{2-}$ on reaction with water[5], [Eq. (30)].

In acetonitrile, $Rh_4(CO)_{12}$ is quantitatively oxidized by H_2SO_4 according to the reaction:[61]

$$Rh_4(CO)_{12} + 4H_2SO_4 \xrightarrow[CH_3CN]{25\,^\circ C} 4[Rh(CH_3CN)_2(CO)_2]HSO_4 + 2H_2 + 4CO \quad (25)$$

This should be contrasted with the protonation of both $Ir_4(CO)_{12}$ [160] and $Ir_4(CO)_8L_4$ [46] by strong acids (H_2SO_4 and CF_3COOH respectively) to give presumably the dications $[Ir_4(CO)_8L_4H_2]^{2+}$ (L = CO, PR_3).

C. Substitution

In the tetranuclear clusters substitution reactions are complicated by the large number of different possible isomers. For instance, in the structure of $Co_4(CO)_{12}$ there is the possibility of three different monosubstituted derivatives (apical, equatorial or axial), and seven different di-substituted isomers. It seems possible that the extensive use of ^{13}C, and where appropriate ^{31}P NMR [66] could make an important contribution to this area.

The most simple substitution is that with ^{13}CO. The following compounds are known to exchange easily (in solution 25 °C, 1 atm): $Co_4(CO)_{12}$ [47, 69], $Co_4(CO)_{11}[P(OMe)_3]$ [66], $Co_3Rh(CO)_{12}$ [141], $Rh_4(CO)_{12}$ [73] and $[Rh_4(CO)_{11}(COOMe)]^-$ [111]. Slow exchange is observed for $[Fe_4(CO)_{13}]^{2-}$ [111] and $FeCo_3(CO)_{12}H$ [69].

The only substitution in a tetranuclear cluster which has been studied kinetically is the progressive substitution in $Ir_4(CO)_{12}$ by triphenylphosphine, and these results have not yet been published in full. The reported experimental data are the following[149]:

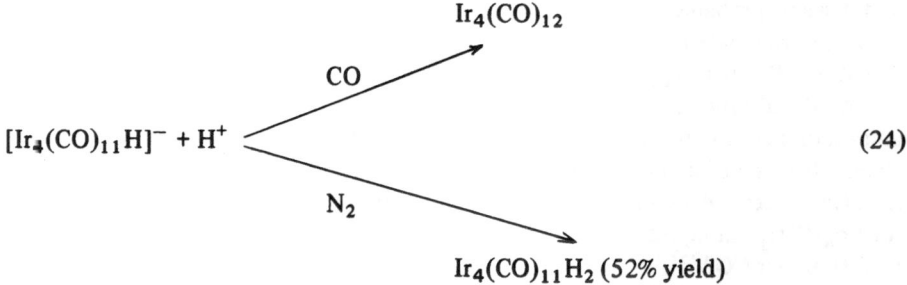

$$Ir_4(CO)_{12} \xrightarrow[106\,^\circ C]{PPh_3} Ir_4(CO)_{11}(PPh_3) \xrightarrow[75\,^\circ C]{PPh_3} Ir_4(CO)_{10}(PPh_3)_2 \xrightarrow[75\,^\circ C]{PPh_3} Ir_4(CO)_9(PPh_3)_3$$

Mechanism	S_N2	S_N1	S_N1	
ΔH^*kcal mol^{-1}	20.6	31.8	31	
ΔS^* e.u.	−22	14	17	(26)
Relative rates of substitution (at 75 °C)	1	30	920	

This data shows that increasing substitution results in an increase of both the S_N1 contribution and the rate of substitution. Both these effects have been ascribed to the formation of bridging carbonyl groups, [Eq. (17)][149].

Unfortunately, several less direct observations indicate that the kinetic conclusions found in the substitution of $Ir_4(CO)_{12}$ are not easily generalized. For instance, it is known that progressive substitution in other clusters, such as $Ru_3(CO)_{10}(NO)_2$[205] and $Co_3(CO)_9CR$[44], always exhibit predominantly S_N1 kinetics. Moreover, it is also known that there is no large increase in the relative rates of substitution in either $Co_4(CO)_{12}$[48] or $Rh_4(CO)_{12}$[61, 250] since both these clusters react with a stoichiometric amount of triphenylphosphine to give essentially the monosubstituted cluster.

In the tetranuclear clusters significant steric control of substitution with tertiary phosphines and phosphites is generally only observed in the fourth substitution step. The synthesis of tetra-substituted derivatives, therefore, requires more severe conditions[22, 46, 86, 248] and then breaking of the cluster can become important. In agreement with the relative trends in the metal-metal bond energies, the ease of synthesis of the tetra-substituted derivatives decreases in the order

$$Ir_4 > Rh_4 > Co_4$$

The syntheses of highly substituted derivatives is possible with phosphites, [*e.g.* $P(OMe)_3$ or $P(OPh)_3$], and chelating bis-phosphines such as DPE[219] and DME[74]. Although the formation of such species might be associated with particular steric effects, it seems more probable that phosphite substitution is favoured by their low basicity and that the chelating phosphines introduce a cage effect, which makes it more difficult to break the cluster. Using *iso*-nitriles which are in principle less sterically demanding, it has been possible to obtain the penta-substituted cluster, $Co_4(CO)_7(tBuNC)_5$ [202].

Substitution with unsaturated hydrocarbons seems to require the prior formation of vacant coordination sites, and the direct reaction with arenes requires particularly severe conditions. This effect is illustrated in the following comparison:

$$Co_4(CO)_{12} \xrightarrow{C_2Et_2,\ \text{benzene},\ 25\,^\circ C,\ 3\ \text{days}} Co_4(CO)_{10}(C_2Et_2)\ (\text{yield } 40\%)^{167)}$$

$$Co_4(CO)_{12} \xrightarrow{\text{mes},\ 60\text{–}70\,^\circ C,\ 6\ h} Co_4(CO)_9(\text{mes})\ (\text{yield } 40\%)^{151)} \qquad (27)$$

IX. Clusters of Cr, Mo, W

In the chromium triad, the only claimed tetranuclear compound is the anion $[Mo_4(CO)_{11}]^{2-}$ which was mentioned together with $[Mo_3(CO)_{14}]^{2-}$, in a study of the reaction of sodium tetrahydroborate and hexacarbonyl-molybdenum in liquid ammonia[14]. Later work failed to confirm the existence of the trinuclear anions, $[M_3(CO)_{14}]^{2-}$ (M = Cr, Mo, W)[109, 150, 169], and, therefore, the real existence of the tetranuclear anion also seems dubious.

The reaction of *trans*-$MCl_2(PPh_3)_2$ (M = Pt [30] and Pd [81]) with $Na[Mo(CO)_3Cp]$ gives substances formulated as $Mo_2M_2(CO)_9Cp_2(PPh_3)_2$, but which are not yet completely characterized. The related complexes $[M_2Fe_2(CO)_{10}Cp_2]^{2-}$ (M = Mo or W) have been obtained on mixing THF solutions of $Fe_2(CO)_9$ and $[M(CO)_3Cp]^-$ at room temperature[130]. The maroon molybdenum complex and red-violet tungsten complex are both very sensitive to oxidation. Their infrared spectra show no bridging carbonyls but their structures have not yet been determined.

X. Clusters of Mn, Tc, Re

Because of the large number of carbonyl groups required to satisfy the noble gas rule and the relatively small metallic radius, serious steric limitations are present for manganese tetranuclear closed clusters[12]. The only known tetranuclear compounds are some mixed clusters, $MnOs_3(CO)_{16}H$ and $MnOs_3(CO)_{13}H_3$ (see later), which have been prepared from $Os_3(CO)_{12}$ and $[Mn(CO)_5]^-$ [161].

The chemistry of the technetium carbonyls has not been extensively studied and tetranuclear species have not yet been reported.

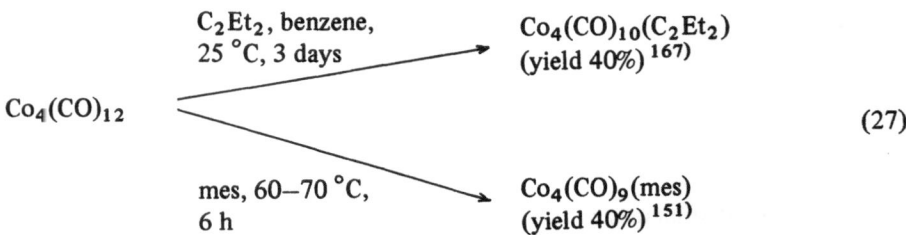

The existence of the neutral rhenium carbonyl $[Re(CO)_4]_n$ was first claimed in 1965 [206] but, although it is easily sublimed, it has not yet been characterized by mass spectrometry and the value of n is still not known. This colourless substance [ν(CO) 2055 and 1995 cm^{-1} in $CHCl_3$] has been obtained as a by-product in the synthesis of $Re_2(CO)_{10}$ starting from Re_2S_7, copper powder, and carbon monoxide at 85 atm, 200 °C[206]. There has also been a report of the compound $Re_4(CO)_{10}(PPh_2Me)_6$, which can be considered to be a substitution product of the hypothetical species, $Re_4(CO)_{16}$; it has been obtained by a photochemical reaction between $Re_2(CO)_{10}$ and PPh_2Me[194]. In both cases, and particularly in the phosphine derivative, a tetrahedral structure seems improbable because of steric constraints.

The first well-characterized compound, the deep red hydridoderivative, $Re_4(CO)_{12}H_4$, was originally prepared by pyrolysis:

$$Re_3(CO)_{12}H_3 \xrightarrow[0.5\ h]{decalin,\ 190\ °C} Re_4(CO)_{12}H_4 \quad (yield \simeq 22\%)\ ^{218)} \tag{28}$$

However, as shown in Eq. (3), the yield is increased to 50–60% when working at 150 °C in a hydrogen atmosphere[146]. The molecular weight has been confirmed by mass spectrometry, and the infrared spectrum indicates the presence of only terminal carbonyl groups (see Table 8). The infrared spectrum shows no bands attributable to Re-H stretching vibrations and this has been interpreted as being due to the hydrogens occupying face-bridging positions; these features have recently been confirmed by X-ray analysis[251a]. An electron count shows that in this diamagnetic compound each rhenium atom has 14 electrons instead of the 15 which is required for a tetrahedral structure. As a result, it was proposed that the tetrahedron contains two double Re-Re bonds which are delocalized over the whole metal skeleton. This electronic unsaturation would be in agreement with the high reactivity of the compound both with Lewis bases (such as Et_2O, CH_3CN, PR_3) and with CO. In the last case, the following degradation is observed[163]:

$$Re_4(CO)_{12}H_4 + 5CO \xrightarrow[25\ °C]{1\ atm} Re_3(CO)_{12}H_3 + Re(CO)_5H \tag{29}$$

Reduction of $Re_2(CO)_{10}$ with sodium tetrahydroborate is an extremely complicated reaction, which is well exemplified by the presence of at least 10 different signals due to Re-H bonds in the ^1H NMR spectrum[96, 147]. Initially a red solution is formed which slowly changes to pale yellow. From the red solution the deep red dianion $[Re_4(CO)_{16}]^{2-}$ (structure in Fig. 5) has been isolated as the tetrabutylammonium salt in 20% yield[11]. From the pale yellow solution the anions $[Re_3(CO)_{12}H_2]^-$, $[Re_3(CO)_{12}H]^{2-}$ and $[Re_4(CO)_{12}H_6]^{2-}$ have been isolated[148]. The structure of the last pale yellow anion has already been discussed in Section V[145]. It is interesting to note that this anion is formally the dimer of the dinuclear monoanion $[Re_2(CO)_6(\mu_2\text{-}H)_3]^-$ [102] and that is has also been prepared by the reaction of $NaBH_4$ on $Re_4(CO)_{12}H_4$ [218]; the corresponding deuteride, $[Re_4(CO)_{12}D_6]^{2-}$ has also been prepared[145].

Although the reaction of methanolic potassium hydroxide (100 °C, 6 h) with $Re_2(CO)_{10}$ has been reported to give the anion $[Re_2(CO)_8O_2H]^-$ [117] recent workers have isolated (65 °C, 15 min) a different anion which has been formulated as $[Re_4(CO)_{16}(OMe)H_4]^{3-}$ [5]. Apparently, the hydrolysis of this colourless anion:

$$[Re_4(CO)_{16}(OMe)H_4]^{3-} + H_2O \longrightarrow [Re_4(CO)_{15}H_4]^{2-} + MeOH + OH^- + CO \quad (30)$$

gives the interesting yellow dianion $[Re_4(CO)_{15}H_4]^{2-}$ which contains the unusual structure shown in Fig. 5.

The following reactions have been reported[108]:

$$Re_2(CO)_{10} + In \xrightarrow[\text{xylene}]{170\,°C-185\,°C} Re_2(CO)_8[\mu_2\text{-}InRe(CO)_5]_2 \ \text{(A) (yield 11\%)} \quad (31)$$

$$Re_2(CO)_{10} + In \xrightarrow[\text{xylene}]{220\,°C-230\,°C} Re_4(CO)_{12}[\mu_3\text{-}InRe(CO)_5]_4 \ \text{(B) (yield 41\%)} \quad (32)$$

It seems probable that the coordinatively unsaturated species, $InRe(CO)_5$, is involved in the synthesis of (A), and that the second species, (B), results from further condensation of (A). The structure of the analogous compound, $Mn_2(CO)_8[\mu_2\text{-}InMn(CO)_5]_2$, is known[212] and it is worth noting that this manganese compound does not undergo a further condensation, corresponding to Eq. (32). Compound (B) has a remarkable thermal stability and decomposes over 296 °C; it is stable in air and also in solution[108]. Its structure is shown in Fig. 6.

Several other examples of compounds of the type $Re_4(CO)_{12}(\mu_3\text{-}E)_4$ are known (E = OH[113], OD[113], OMe[113], Cl[80], SMe[106], SPh[1], SePh[1]). However, in all these compounds each hetero atom, E, donates 5 electrons and this results in the disappearance of all M-M bonds as shown by X-ray analysis of $[Re_4(CO)_{12}(\mu_3\text{-}SMe)_4]$, ($d_{Re\text{-}Re} = 3.853 - 3.957$, $\bar{d}_{Re\text{-}C} = 1.87(5)$, $\bar{d}_{C\text{-}O} = 1.20(5)$ Å)[106].

The anion $[Re(CO)_5]^-$ reacts readily with $Ru_3(CO)_{12}$ to give a complicated mixture of products. The following scheme shows only the mixed compounds which have been isolated[161]:

$$Ru_3(CO)_{12} + [Re(CO)_5]^- \xrightarrow[\text{THF}]{20\,°C,\ 20\ min} \underset{\text{(orange)}}{[ReRu_3(CO)_{16}]^-} \xrightarrow{+H^+} \underset{\text{(pale yellow)}}{Re_2Ru(CO)_{12}H_2} + \underset{\text{(yellow)}}{Re_2Ru_2(CO)_{16}H_2} \quad (33)$$

The corresponding reaction with $Os_3(CO)_{12}$ requires more severe conditions. Again the reaction is extremely complicated but some control of the nature of the products can be obtained by varying the temperature and reaction time. The following scheme shows only the products which have been isolated (by TLC) after acidification[161]:

$$Os_3(CO)_{12} + [Re(CO)_5]^- \xrightarrow[\text{reflux}]{\text{THF}}$$

i) 10 min
ii) H^+ → $ReOs_3(CO)_{15}H$ (orange)

i) 20 h
ii) H^+ → $ReOs_3(CO)_{13}H_3$ (red)

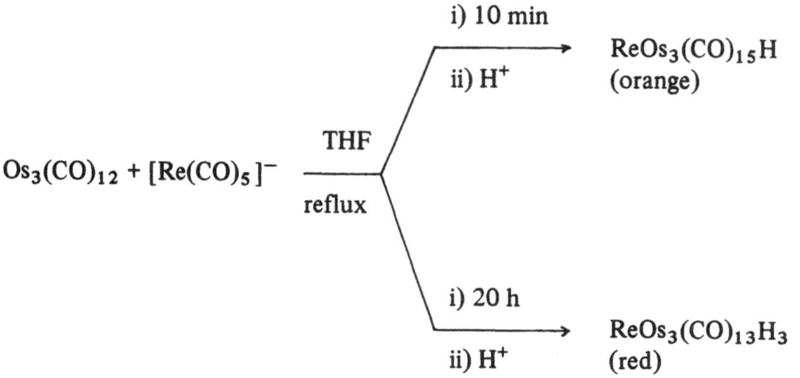

(34)

The compounds $MOs_3(CO)_{13}H_3$ (M = Mn, Re) are clearly analogous to the carbonyl hydrides of the iron triad and are more conveniently discussed in the next section. The clusters $[ReRu_3(CO)_{16}]^-$, $Re_2Ru_2(CO)_{16}H_2$, $MnOs_3(CO)_{16}H$ and $ReOs_3(CO)_{16}H$ (not isolated in a pure state) are all isoelectronic. The two different possible structures which have been proposed are a square of metal atoms and a structure of the type 3 + 1, analogous to the structure subsequently found for the dianion $[Re_4(CO)_{15}H_4]^{2-}$, (Fig. 5a). In the case of $ReOs_3(CO)_{15}H$ a structure analogous to that of the dianion $[Re_4(CO)_{16}]^{2-}$ (Fig. 5b) has been proposed. The infrared spectra of $Re_2Ru_2(CO)_{16}H_2$ and $MOs_3(CO)_{16}H$ (M = Mn or Re) show only terminal carbonyls to be present, and mass spectra agree with the proposed formulations; their ^1H NMR spectra have not been reported[161].

XI. Clusters of Fe, Ru, Os

Despite the early discovery of $[Fe_4(CO)_{13}]^{2-}$,[119] inspection of Table 1 shows that the majority of tetranuclear clusters of this trial are formed by ruthenium.

A. Hydridocarbonyls and Their Derivatives

The synthesis of $[Fe_4(CO)_{13}]^{2-}$ (structure in Fig. 10) has been extensively studied by Hieber and co-workers. The reaction of a variety of ligands[f] with $Fe(CO)_5$ results in redox condensation reactions and formation of the dark red dianion $[Fe_4(CO)_{13}]^{2-}$

---

[f] py[119, 120], lutidine[121], α-PE[121], iso-quin[121, 122], py-N-oxide[123], pyrrolidone[121, 122], N-methylpyrrolidone[121, 122], pyrrolidine[121], morpholine[121], methylamine[121], DEF[121], DMF[121], dmso[123], quin[122].

which can also be prepared in a similar way by addition of nitrogen bases (py[120], py-N-oxide[123], phen in the presence of pyridine[120],) to $Fe_3(CO)_{12}$. The best preparative route is the reversible reaction[119]:

$$5\,Fe(CO)_5 + 6\,py \underset{60\,°C}{\overset{80\,°C}{\rightleftharpoons}} [Fepy_6][Fe_4(CO)_{13}] + 12\,CO \text{ (yield} \sim 80\%) \qquad (35)$$

The formation of $[Fe_4(CO)_{13}]^{2-}$ in these systems involves a complicated series of reactions but the final condensation step is possibly related to the following reaction[127]:

$$[Fepy_6][Fe_3(CO)_{11}] + Fe(CO)_5 \longrightarrow [Fepy_6][Fe_4(CO)_{13}] + 3\,CO \qquad (36)$$

The same dianion has been obtained in 90% yield using $Cr(\eta^6\text{-Ph-Ph})_2$ as a reducing agent

$$2\,Cr(\eta^6\text{-Ph-Ph})_2 + 4\,Fe(CO)_5 \xrightarrow[90\,°C]{benzene} [Cr(\eta^6\text{-Ph-Ph})_2]_2[Fe_4(CO)_{13}] + 7\,CO \qquad (37)$$

The strongest carbonyl band (1955 cm^{-1}) in a methanol solution of $[Fe_4(CO)_{13}]^{2-}$ shifts to 1990–2018 cm^{-1} on addition of excess dilute acid[171] due to the formation of $[H(solvent)_x][Fe_4(CO)_{13}H]$ (structure in Fig. 18) which, contrary to previous work[118, 119], does not easily transform into $Fe_4(CO)_{13}H_2$[171]. In fact, $[H(solvent)_x][Fe_4(CO)_{13}H]$ is a very strong acid and requires dissolution in non-polar solvents or alternatively addition of concentrated sulphuric acid in order to convert it into the neutral dihydrido derivative, $Fe_4(CO)_{13}H_2$. This brown-black pyrophoric dihydride, which gives a deep red solution in ether, is extremely unstable and readily decomposes both in the presence and absence of hydrogen to give $Fe_3(CO)_{12}$[172].

The related monoanion $[Fe_3Co(CO)_{13}]^-$ has been obtained [Eq. (10)] although it has not yet been completely characterized[61].

The instability of $Fe_4(CO)_{13}H_2$ should be contrasted with the stability of $M_4(CO)_{13}H_2$, $FeM_3(CO)_{13}H_2$ (M = Ru, Os) and provides another example of the general increase in thermal stability of hydrido derivatives with the heavier metals[147]. However, the preparative routes to the ruthenium and osmium dihydrido derivatives (Tables 1 and 2) and the homologous trihydrides, $MOs_3(CO)_{13}H_3$ (M = Mn or Re) are generally not well developed, and it is worthwhile noting that the dianions, $[M_4(CO)_{13}]^{2-}$ (M = Ru, Os) have not yet been isolated.

$Ru_4(CO)_{13}H_2$ has been obtained from $Ru_3(CO)_{12}$ (by refluxing in various solvents[140, 137] or by acidification of the solution obtained after reaction with various reducing agents such as sodium tetrahydroborate in THF[138]) and from $[(C_6H_6)RuCl_2]_n$ by reaction with zinc in ethanolic lithium acetate solution[155]. Two isomers of this compound (designated α and β) have been claimed (see Table 10) although most preparations usually give the α-isomer which has the structure shown in Fig. 9. The osmium dihydride, $Os_4(CO)_{13}H_2$, which is probably *iso*-structural with α-$Ru_4(CO)_{13}H_2$, has been obtained from the mixture resulting on pyrolysis of $Os_3(CO)_{12}$ at 230° in the presence of traces of water[87] and by acidification of the solution obtained on reduction of $Os_3(CO)_{12}$ with alkali hydroxide in alcohol[139].

The mixed carbonyls, $FeM_3(CO)_{13}H_2$ (M = Ru[158, 252], Os[197]) have been prepared in low yields as shown in Eqs. (38) and (39):

$$\left.\begin{array}{l} Ru_3(CO)_{12} + Fe(CO)_5 \xrightarrow[24\,h]{100\,^\circ C/\text{trace } H_2O} \\ \\ [Ru(CO)_3Cl]_2 + Fe(CO)_5 \xrightarrow{100\,^\circ C,\ 30\,h} \end{array}\right\} \longrightarrow FeRu_3(CO)_{13}H_2 \qquad (38)$$

$$Os(CO)_4H_2 + Fe_2(CO)_9 \xrightarrow[3\,h]{25\,^\circ C} \underset{(6\%\ \text{yield})}{FeOs_3(CO)_{13}H_2} \qquad (39)$$

The structure of $FeRu_3(CO)_{13}H_2$ is shown in Fig. 10 and the infrared spectrum of the iron/osmium analogue suggests it is *iso*-structural.

The homologous clusters, $MOs_3(CO)_{13}H_3$ (M = Mn[161] or Re[161]), have been prepared, as shown in Eq. (34). They have both been characterized by mass spectrometry and have been suggested to have a structure similar to that of $[Fe_4(CO)_{13}]^{2-}$ (Fig. 10) although the lowest reported carbonyl band (M = Mn, 1830 cm^{-1}; M = Re, 1900 cm^{-1}) seems rather high for a face-bridging carbonyl group. No ^1H NMR data have been reported.

Refluxing $Ru_3(CO)_{12}$ in octane for 1 h in the presence of hydrogen [(Eq. (14)] provides the best route to $Ru_4(CO)_{12}H_4$ [163], which is obtained as a yellow powder in 88% yield. The deuteride, $Ru_4(CO)_{12}D_4$ has been prepared using deuterium instead of hydrogen.

The reaction of hydrogen with the dihydro derivatives, $Ru_4(CO)_{13}H_2$ [146, 163] and $FeRu_3(CO)_{13}H_2$ [163] occurs readily with elimination of one carbonyl group and concomitant relief of steric-crowding to form the tetrahydrido derivatives $Ru_4(CO)_{12}H_4$ and $FeRu_3(CO)_{12}H_4$ respectively. The reaction of $Os_3(CO)_{12}$ with hydrogen is less facile and gives only a 29% yield of the colourless tetrahydride, $Os_4(CO)_{12}H_4$, after refluxing in octane for 41 h [Eq. (15)][163]. Nevertheless this represents a considerable improvement in yield over other methods[87, 139, 196]. Hydrogen abstraction from $M_4(CO)_{12}H_4$ with reformation of $M_4(CO)_{13}H_2$ is not readily accomplished, although the reaction of either cyclo-octene[40] or ethylene[41] with $Ru_4(CO)_{12}H_4$ under a nitrogen atmosphere did give α-$Ru_4(CO)_{13}H_2$ (32% and 30% yield respectively) together with some triruthenium species.

The previously reported[137, 140] existence of α- and β-isomers of $Ru_4(CO)_{12}H_4$ has recently been disputed by Kaesz *et al.*[163] who only find evidence for the α-isomer. Both Kaesz *et al.*[163] and Piacenti *et al.*[210] have obtained a sample of $Ru_4(CO)_{12}H_4$ of improved purity over the earlier preparations[138, 140] and the infrared spectrum consists of only five carbonyl bands consistent with the D_{2d} structure

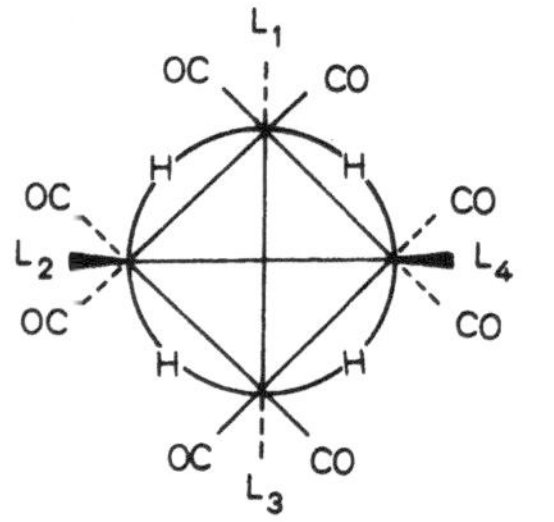

Fig. 22. Schematic representation of the structure of $Ru_4(CO)_8L_4H_4$, (L_{1-4} = CO or PR_3)

(Fig. 22, L_1 .... $_4$ = CO). The infrared and ^1H NMR spectra of $Os_4(CO)_{12}H_4$ suggest a similar D_{2d} structure, (see Table 10).

The reduction of $Ru_4(CO)_{12}H_4$ in ethanolic KOH, followed by addition of tetraphenylarsonium chloride, gives orange-red crystals of $(Ph_4As)[Ru_4(CO)_{12}H_3]$[164], [Eq. (40)]:

$$Ru_4(CO)_{12}H_4 \xrightarrow[2)\ +\ Ph_4AsCl]{1)\ KOH/EtOH/55\ ^\circ C} (Ph_4As)[Ru_4(CO)_{12}H_3] \quad (84\%\ \text{yield}) \tag{40}$$

This first well authenticated tetraruthenium anion is an important addition to this series of tetraruthenium compounds. The ^1H NMR spectrum at room temperature shows a single resonance at 26.9 τ which becomes more complicated at lower temperatures due to the formation of isomers[164], (see Table 10 and Section VIB).

The related tetranuclear cluster $Os_2Co_2(CO)_{12}H_2$, has been prepared, [Eq. (41)]:

$$Os(CO)_4H_2 + Co_2(CO)_8 \xrightarrow[heptane]{25\ ^\circ C,\ 5\,h} + OsCo_2(CO)_{11} + Os_2Co_2(CO)_{12}H_2 \quad (\text{yield } ca.\ 10\%) \tag{41}$$

It is an orange solid which decomposes > 130 °C and has no infrared absorptions due to bridging carbonyls[197].

B. Substituted Ru_4 Clusters

The reaction of $Ru_4(CO)_{12}H_4$ with phosphites or tertiary phosphines results in substitution of up to four carbonyl groups to give clusters of the type $Ru_4(CO)_{12-n}L_nH_4$ [L = $P(OMe)_3$, n = 1,2,3 or 4[162]; L = PBu_3, n = 1,3 or 4[210]; L = PPh_3, n = 1,2,3 or 4[210]]. The degree of substitution can generally be controlled by a suitable choice of conditions, but the lower substituted derivatives are better prepared by the disproportionation reaction[210]:

$$Ru_4(CO)_{12}H_4 + Ru_4(CO)_{10}(PPh_3)_2H_4 \xrightarrow[100\ ^\circ C]{H_2\ \text{pressure}} Ru_4(CO)_{11}(PPh_3)H_4 \tag{42}$$

$Ru_4(CO)_{12-n}(PPh_3)_nH_4$ (n = 2 or 3) is also formed on reaction of $Ru(CO)_3(PPh_3)_3$ with hydrogen under pressure[210], and $Ru_4(CO)_9[P(OPh)_3]_3H_4$ is formed, together with other metallation products, on heating $Ru_3(CO)_9[P(OPh)_3]_3$ in decalin[36]. All these complexes have the structure shown in Fig. 22 with $L_{1...4}$ = Co or PR_3. Only one hydride resonance is observed for each value of *n*, even at $-100\ ^\circ C$, and equal coupling to the phosphorus atoms is observed due to intramolecular exchange, (see Section VIB).

The reaction of either α-$Ru_4(CO)_{13}H_2$ or $Ru_4(CO)_{12}H_4$ with cyclooctadienes[40] in refluxing cyclohexane gives the deeply coloured derivatives $Ru_4(CO)_{12}L$ ($L = C_8H_{10}$ or C_8H_{12}), $Ru_4(CO)_{11}(C_8H_{10})$ as well as species which result from degradation of the cluster[40]. The X-ray structure of $Ru_4(CO)_{11}(C_8H_{10})$ has been determined[186] (Fig. 17) and the other two analogues probably have similar structures with the additional carbonyl group displacing the olefinic group.

The reaction of $Ru_3(CO)_{12}$ with acetylenes in refluxing hexane gives a 5–10% yield of brown $Ru_4(CO)_{12}(RC_2R')$ (R = Ph, R' = Ph, Me or Et), which probably have structures related to that of $Co_4(CO)_{10}(C_2R_2)$ (Fig. 17), and can be converted to $[Ru_4(CO)_{12}(RC_2R')H]^+$ on addition of strong acid[142]. Further carbonyl substitution occurs in $Ru_4(CO)_{12}(RC_2R')$ on refluxing with the appropriate acetylene to give orange $Ru_4(CO)_{11}(RC_2R')(PhC_2R'')$ (R'' = Me, Et)[142], which probably contain a metallocycle. The formation of these clusters is unusual since predominantly trinuclear species are formed on reaction of $M_3(CO)_{12}$ (M = Fe[39], Os[136]) with acetylenes. The infrared spectra of all the tetranuclear ruthenium derivatives show only terminal carbonyl bands; the hydride resonance of $[Ru_4(CO)_{12}(PhC_2Ph)H]^+$ (33.4 τ) shows little change down to $-60\ ^\circ C$. Hydrogenation of $Ru_4(CO)_{12}(C_2Ph_2)$ in *n*-heptane at reflux gives an almost quantitative yield of $Ru_4(CO)_{12}H_4$ and *trans*-stilbene[142].

The reaction of $Ru_3(CO)_{12}$ with cyclododeca-1,5,9-trienes gives a mixture of triruthenium clusters and $Ru_4(CO)_{10}(C_{12}H_{16})$ which have been separated by chromatography[33]. The X-ray structure of this tetraruthenium cyclododecatrienyl compound is shown in Fig. 17.

Prolonged reflux (96h.) of 4,6,8-trimethylazulene, (L), with $Ru_3(CO)_{12}$ in ligroin gives $Ru_6(CO)_{17}C$ and $Ru_4(CO)_9L$[62], which has the structure shown in Fig. 16.

C. Clusters Containing the η^5-Cyclopentadienyl Group

Prolonged (12 days) reflux of $[Fe(CO)_2Cp]_2$ in xylene gives the dark green diamagnetic compound $[Fe(CO)Cp]_4$ [154], which has the structure shown in Fig. 7[201]. This neutral cluster undergoes ready oxidation with bromine to give the black paramagnetic (2.13 BM) cation, $[Fe(CO)Cp]_4^+$ [154] which has been isolated with the following anions, Cl^-, Br_3^-, I_5^-, PF_6^-. The crystal structure of the hexafluorophosphate salt has been determined and is shown in Fig. 7[234]. The electrochemical behaviour[94] [Eqs. (19) and (23)] has already been referred to and it is worth noting that the most oxidized and reduced species, $[Fe(CO)Cp]_4{}^n$ ($n = +2$ and -1 respectively), detected in this study have not yet been isolated.

The face-bridging carbonyls in $Fe_4(CO)_4Cp_4$ are strongly basic and, with Lewis acids, form a number of *O*-bonded adducts of the type $Fe_4(CO)_{4-x}(COA)_xCp_4$

($A = BF_3, x = 1, 2$ or 4; $A = BCl_3$ or $BBr_3, x = 1$ or 2; $A = AlBr_3, x = 1,2,3$ or 4[166]; $A = AlEt_3, x = 4$[7]). Infrared spectra of the adducts show a shift of the carbonyl bands to lower frequency [Eq. (16)] and the number of carbonyl bands observed is consistent with the expected symmetry of the adduct[166].

The kinetics of the redox reaction:

$$2[Fe_4(CO)_4Cp_4]^+ + Fe_2(CO)_4Cp_2 \xrightarrow{CH_3CN} 2[Fe(CO)_2Cp(CH_3CN)]^+ + \\ + 2Fe_4(CO)_4Cp_4 \qquad (43)$$

have been studied and ΔH^{+} is very small (1.8 ± 0.2 kcal mol^{-1})[28].

Controlled pyrolysis of $[Ru(CO)_2Cp]_2$ gives the deep purple tetraruthenium compound, $[Ru(CO)Cp]_4$, analogous to the iron derivative discussed above, together with small amounts of $Ru_4(CO)_6Cp_3$[18]. The infrared spectrum of $[Ru(CO)Cp]_4$ has a band at 1616 cm^{-1} due to the face-bridging carbonyl group and the mass spectrum suggests that it is more stable than the iron analogue. However, there was no evidence for the formation of mixed iron/ruthenium clusters.

$Ru_4(CO)_6Cp_3$ has not been well-characterized due to its low solubility and isolation only in small amounts. However, a parent ion was obtained in the mass spectrum and the structure which has been suggested is shown in Fig. 23.
The related complexes $[M_2Fe_2(CO)_{10}Cp]^{2-}$ (M = Mo or W) have already been mentioned (Section IX)[130].

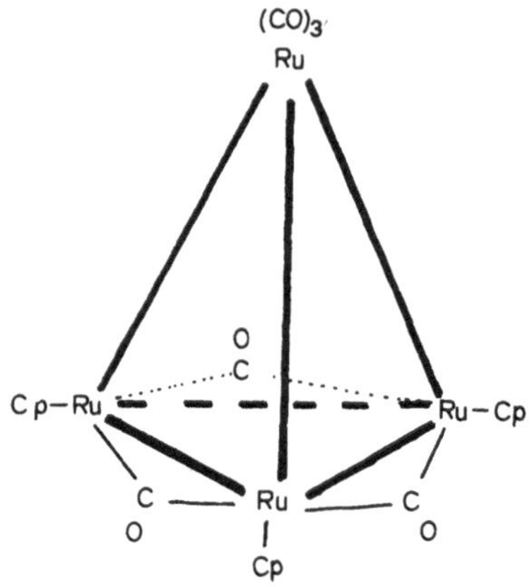

Fig. 23. Suggested structure of $Ru_4(CO)_6Cp_3$

D. Miscellaneous

The reaction of $Os_3(CO)_{12}$ with $Au(PPh_3)X$ (X = Cl, Br, SCN, I) in refluying xylene for 10–30 min gives the red clusters, $Os_3(CO)_{10}[AuPPh_3]X$[26]; the yields decrease in the order Cl > Br > SCN > I. The analogous compound $Os_3(CO)_{10}[AuP(C_6H_4Me)_3]Cl$ has also been prepared. They have all been characterized by mass spectrometry and (for X = Cl, Br) by X-ray structural analysis[99] (Fig. 20). A comprehensive infrared/Raman study has been carried out[26].

The reaction of $Ru_3(CO)_{12}$ with nitric oxide in refluxing benzene gives mainly $Ru_3(CO)_{10}(NO)_2$ but after chromatography 0.5% of a pale yellow compound was obtained[204]. The exact formulation of this compound is not yet known but a parent ion corresponding to $Ru_4(CO)_{12}(N_2O)^+$ or $Ru_4(CO)_{11}(NO)(NCO)^+$ was obtained. The infrared spectrum showed a band at 1510 cm^{-1} due to $\nu(NO)$ and no bridging carbonyl groups[204].

The white tetranuclear osmium compound $Os_4(CO)_{12}O_4$ has been isolated on reaction of OsO_4 in xylene with CO at 175 °C and 175 atm for 24 h[27]. X-ray analysis shows a cubane structure with oxygen atoms bridging a face by donating 4 electrons, resulting in there being no osmium-osmium bonds[27], ($d_{Os\text{-}Os}$ = = 3.190–3.253 Å; $\bar{d}_{Os\text{-}CO}$ = 1.915 Å)[31].

XII. Clusters of Co, Rh, Ir

A. Synthesis of the Dodecacarbonyls

As is usual for carbonyl clusters containing first row elements the synthesis of $Co_4(CO)_{12}$ starts from the simpler carbonyl, $Co_2(CO)_8$:

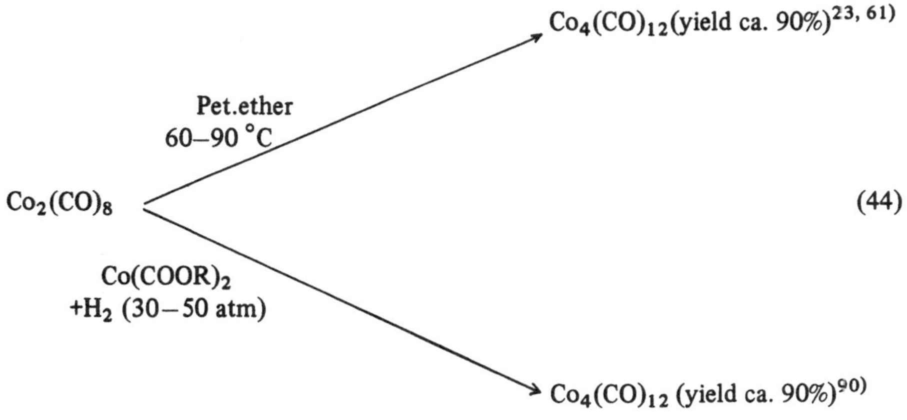

(44)

The formation of $Co_4(CO)_{12}$ from $Co_2(CO)_8$ is easily reversed both under pressure (27–110 atm; 35–105 °C)[24] and at atmospheric pressure (*iso*-propanol in the presence of a trace of halide[61]). Such ready reversibility shows that the thermodynamically more stable product (1 atm CO, 25 °C) is $Co_2(CO)_8$.

This should be contrasted with the synthesis of $Rh_4(CO)_{12}$ which, owing to the low stability of $Rh_2(CO)_8$ [107, 250], can be carried out directly under a pressure of carbon monoxide starting with a Rh(III) or Rh(I) halide, in the presence of a halide acceptor (such as copper) working at temperatures lower than 80 °C[21, 49, 115]. In each case, intermediate formation of Rh(I) derivatives is involved and the synthesis of $Rh_4(CO)_{12}$ can be more easily carried out at atmospheric pressure, using the two reactions shown in the following equations:

$$2RhCl_3 \cdot (H_2O)_x + 6CO \xrightarrow[1\ atm]{90\ °C} Rh_2(CO)_4Cl_2 + 2COCl_2 + 2xH_2O \quad (45)$$

(yield ca. 90%)[190]

$$2Rh_2(CO)_4Cl_2 + 6CO + 4NaHCO_3 \xrightarrow[wet\ hexane]{25\ °C,\ 1\ atm} Rh_4(CO)_{12} + 6CO_2 + 4NaCl + 2H_2O \quad (46)$$

(yield ca. 80–85%)[45, 57]

Alternatively, the two steps can be carried out *in situ*, controlling the pH by addition of disodium citrate:

$$K_3[RhCl_6] + 2Cu + 4CO \xrightarrow[H_2O,\ \sim 12h]{25\ °C,\ 1\ atm} K[Rh(CO)_2Cl_2] + 2CuCl \cdot CO + 2KCl \quad (47)$$

$$4K[Rh(CO)_2Cl_2] + 6CO + 2H_2O \xrightarrow[H_2O,\ pH\ 4]{25\ °C,\ 1\ atm\ \sim 12h.} Rh_4(CO)_{12} + 2CO_2 + 4HCl + 4KCl \quad (48)$$

(yield ca. 85–95%)[183]

The synthesis of $Ir_4(CO)_{12}$ is similar to that of $Rh_4(CO)_{12}$, except that, in this case, the reactions can be carried out not only under high pressure (350 atm) but also at high temperatures (100–200°)[114]. An infrared study, under pressure (100 atm, 100°) using copper bronze as halide acceptor[249], indicates a reduction sequence of the following type:

$$[Ir^{IV}Cl_6]^{2-} \longrightarrow [Ir^{III}(CO)Cl_4 \cdot ROH]^- \longrightarrow [Ir^{III}(CO)_2Cl_4]^- \longrightarrow [Ir^{I}(CO)_2Cl_2]^- \longrightarrow Ir_4(CO)_{12} \quad (49)$$

(yield 70–80%)[249]

This synthesis has also been carried out at 60 °C and 40 atm in methanol solution containing sodium bicarbonate[49]. A route starting from preformed $Ir(CO)_2Cl$(*p*-toluidine) and working at 90 °C and 4 atm has been described (yield 80–85%)[228] and finally the synthesis can also be carried out slowly at atmospheric pressure working in the presence of iodide (yield 74–78%, ca. 2 days[176]). $Ir_4(CO)_{12}$ does not react with carbon monoxide up to 550 atm and 200 °C, while with a 1 : 1 mixture of $H_2 : CO$ there is partial transformation into $Ir(CO)_4H$[247].

We have already mentioned in Eq. (7) the synthesis of $Co_2Rh_2(CO)_{12}$. A general method of synthesis consists of the precipitation of the mixed carbonyls from an aqueous solution:

$$2Na[Rh(CO)_2Cl_2] \xrightarrow[H_2O,\ 25\ °C]{2Na[Co(CO)_4]} Co_2Rh_2(CO)_{12} \quad \text{(yield 70–80\%)}^{184)}$$

$$2Na[Rh(CO)_2Cl_2] \xrightarrow[H_2O,\ 25\ °C]{2Na[Ir(CO)_4]} Rh_2Ir_2(CO)_{12} \quad \text{(yield 60–70\%)}^{184)} \quad (50)$$

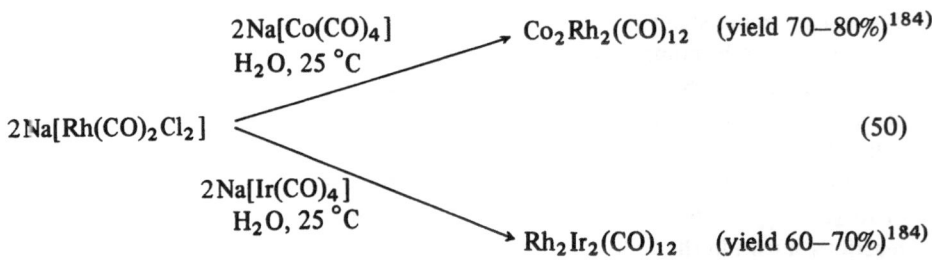

Alternatively the synthesis of these mixed compounds can be carried out in hydrocarbon solvents:

$$Rh_2(CO)_4Cl_2$$

$$\nearrow \quad \begin{array}{l} Zn[Co(CO)_4] \\ \text{toluene, 25 °C} \end{array} \quad \nearrow \quad \begin{array}{l} Co_2Rh_2(CO)_{12} + ZnCl_2 \\ \text{(yield 70–80\%)}^{184)} \end{array}$$

$$\searrow \quad \begin{array}{l} 2\,Tl[Co(CO)_4]_3 \\ \text{hexane} \end{array} \quad \searrow \quad 2\,Co_3Rh(CO)_{12} + 4\,CO + 2\,TlCl^{131)}$$

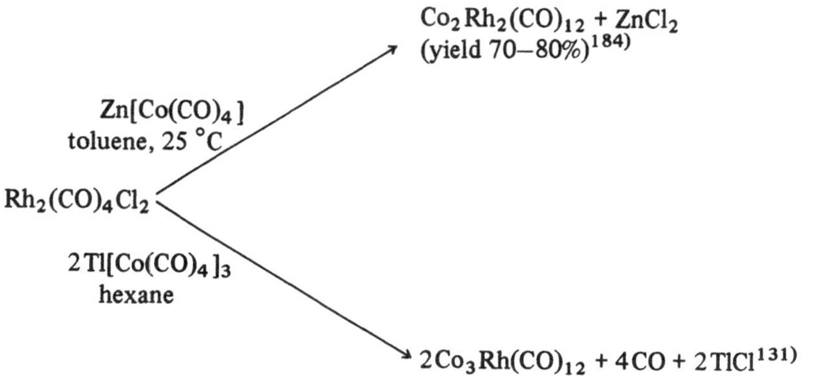

(51)

B. Physical Properties of the Dodecacarbonyls

Some physical and chemical properties of the dodecacarbonyls are tabulated in Table 12.

Table 12. Physical and chemical properties of the dodecacarbonyls $M'_{4-n}M''_n(CO)_{12}$ [184)]

Compound	Colour	λ_{max}[1)] (nm)	Dec. temp. (°C)	Volatility at 0.01 Torr	Reactivity Air	THF	ΔH°_f kcal mol^{-1} [68)]
$Co_4(CO)_{12}$	Dark Brown	375	100	Sub. 57 °C	+	+	-441 ± 4
$Co_3Rh(CO)_{12}$	Brown	365	120 (60)[2)]	Dec.	+	+	
$Co_2Rh_2(CO)_{12}$	Brown	348	120 (60)[2)]	Dec.	+	+	
$Co_2Ir_2(CO)_{12}$	Dark Red	[3)]	190	Sub. 60 °C	$-$[4)]	+	
$Rh_4(CO)_{12}$	Red	300	130 (35)[2)]	Dec.	$-$		-442 ± 3
$Rh_3Ir(CO)_{12}$	Orange	300	130 (85)[2)]	Dec.	$-$	$-$	
$Rh_2Ir_2(CO)_{12}$	Orange	310	150	Dec.	$-$	$-$	
$Ir_4(CO)_{12}$	Yellow	319 [5)]	210	$-$[6)]	$-$	$-$	-435 ± 4

[1)] Hexane solutions.
[2)] Temperatures in parenthesis refer to decomposition in solution to give $M_6(CO)_{16}$ species.
[3)] No observed maximum.
[4)] Stable during several days in the solid state, less in solution.
[5)] Chloroform solution
[6)] Sublimes very slowly at 120–170 °C in a stream of CO.

The only species which can be purified by sublimation are $Co_4(CO)_{12}$ and $Co_2Ir_2(CO)_{12}$. $Rh_4(CO)_{12}$ is readily and irreversibly transformed into $Rh_6(CO)_{16}$ and this is why it must be prepared at temperatures less than 80 °C: all the other rhodium containing species behave similarly and they are all purified by crystallization from hydrocarbon solvents. However, purification of $Ir_4(CO)_{12}$ is more difficult since, although it is stable at high temperatures, it volatilises very slowly and its low solubility in common solvents necessitates a hot solvent extraction procedure (for instance with 1,2-$C_2H_4Cl_2$) [102]. The infrared spectra[184] of all these compounds, with the exception of $Ir_4(CO)_{12}$, suggest that they adopt the icosahedral structure which is known for $Co_4(CO)_{12}$ (Fig. 4), $Rh_4(CO)_{12}$ and $Co_2Ir_2(CO)_{12}$. This hypothesis has been confirmed for $Co_3Rh(CO)_{12}$ by ^{13}C NMR[126].

On descending the sub-group the colour generally becomes lighter, indicating an increased separation between the frontier orbitals, which are believed to be predominantly associated with the metal frame, and therefore a decrease of the metallic character. The overall constancy of the values of the enthalpies of formation of these compounds, and the use of an estimated sublimation enthalpy of ca. 23–25 kcal mol^{-1} [68], gives rise to a net enthalpic term (per mole of CO) associated with the reaction of formation, from M(g) and CO(g), of about –8.5 kcal mol^{-1}.

C. Reduction and Oxidation Reactions

As shown in Table 12 all the dodecacarbonyls containing cobalt are sensitive to Lewis bases, which tend to induce disproportionation. This behaviour is well documented for $Co_2(CO)_8$ [39] but has been much less studied in the case of $Co_4(CO)_{12}$:

$$3Co_4(CO)_{12} + 24B \longrightarrow 4[CoB_6][Co(CO)_4]_2 + 4CO \qquad (52)$$

$$(B = NH_3\ ^{15)}, Py^{116)}, MeOH^{53)}, EtOH^{53)},)$$

It is interesting that $Co_4(CO)_{12}$ is perfectly stable in the more sterically demanding *iso*-propyl[53] and *tert*-butyl alcohols[245]. This disproportionation reaction is unknown for $Rh_4(CO)_{12}$ and $Ir_4(CO)_{12}$, probably because of the reduced thermodynamic stability of the cationic species of these noble metals.

In particular solvents such as diethyl- and di-*iso*-propyl-ether a rapid reaction between $Co_4(CO)_{12}$ and the anion $[Co(CO)_4]^-$ is observed:

$$2[Co(CO)_4]^- + Co_4(CO)_{12} \underset{i\text{-PrOH or THF}}{\overset{Et_2O \text{ or } i\text{-}Pr_2O}{\rightleftharpoons}} 2[Co_3(CO)_{10}]^- \qquad (53)$$

The deep red colour and the characteristic low frequency infrared absorption (~1600 cm^{-1}) of the anion $[Co_3(CO)_{10}]^-$ disappear simply by changing the solvent[61].

In comparison with the remarkable stability of the anions $[Co(CO)_4]^-$, $[Co_6(CO)_{15}]^{2-}$ and $[Co_6(CO)_{14}]^{4-}$, the low stability of the anionic tetranuclear clusters of cobalt seems extraordinary. Thus the low stability of the anion $[Co_4(CO)_{11}(COO\textit{i}\text{-}Pr)]^-$ has prevented its isolation in a pure form[61] and moreover there is, as yet, no evidence for the hypothetical dianion $[Co_4(CO)_{11}]^{2-}$.

The stability of the tetranuclear anions increases on descending the sub-group; we have already shown in Eq. (21) that the orange carbomethoxy species, $[Rh_4(CO)_{11}(COOMe)]^-$, is easily obtained by reaction with sodium methoxide and that it is transformed into $[Rh_4(CO)_{11}]^{2-}$ by further reaction with hydroxide[185]. The mono-anion $[Ir_4(CO)_{11}(COOMe)]^-$ can be quantitatively converted back to $Ir_4(CO)_{12}$ by reaction with acetic acid:

$$[Ir_4(CO)_{11}(COOMe)]^- + H^+ \xrightarrow{CO} Ir_4(CO)_{12} + MeOH \quad (54)$$

This provides one of the few available methods for the purification of $Ir_4(CO)_{12}$ [102].

Reduction of $Ir_4(CO)_{12}$ is a more complicated reaction due to the several tetranuclear species which are formed and to the concurrent catalytic formation of formates[9]. The above reaction with methoxide should be contrasted with the reaction of $Ir_4(CO)_{12}$ with a suspension of sodium carbonate in methanol, which gives the anion, $[Ir_4(CO)_{11}H]^-$ [9]. It seems probable that this anion is formed through formation of a COOH group followed by H-migration.

$$Ir_4(CO)_{12} + OH^- \longrightarrow [Ir_4(CO)_{11}H]^- + CO_2 \quad (55)$$

The anion $[Ir_4(CO)_{11}H]^-$ is stable under a CO atmosphere (*trans*-effect of hydride?), whereas under nitrogen it rapidly gives brown products of higher nucleari-ty, which are not yet well-characterized[9]. The dianion $[Ir_4(CO)_{10}H_2]^{2-}$ has also been recently characterized[102].

The reaction of strong acids with $Co_4(CO)_{12}$ results in demolition and formation *inter alia* of $Co(CO)_4H$[61] while $Rh_4(CO)_{12}$ gives rise to the oxidation already mentioned in Eq. (24) and the more robust $Ir_4(CO)_{12}$ is believed to give tetranuclear cationic species[160].

D. Substituted Dodecacarbonyls and Related Clusters

With tertiary phosphines, phosphites and *iso*-nitriles, the substitution of the carbonyl groups in $Co_4(CO)_{12}$ and $Rh_4(CO)_{12}$ occurs very easily up to the tri-substituted complex. It is sufficient to add slowly at room temperature a solution of the ligand to a solution of the dodecacarbonyl:

$$M_4(CO)_{12} \underset{CO}{\overset{L}{\rightleftharpoons}} M_4(CO)_{11}L \underset{CO?}{\overset{L}{\rightleftharpoons}} M_4(CO)_{10}L_2 \underset{CO?}{\overset{L}{\rightleftharpoons}} M_4(CO)_9L_3 \quad (56)$$

In the case of $Co_4(CO)_{12}$ it has been shown that for $L = AsPh_3$ or $SbPh_3$, the first step of the substitution is reversible[48], although this is not the case for $P(OMe)_3$ [66]. In the case of rhodium, the following reversible equilibrium has been reported[250]:

Table 13. Substituted products of $M_4(CO)_{12}$ with the ligands PR_3, AsR_3, SbR_3, $P(OR)_3$, SR_2 and RNC

$Co_4(CO)_{11}L$	L = RNC (R = Me, Et, *t*Bu, Cy, Bz) [202], PEt_3 [70], $PMePh_2$ [70], PPh_3 [48], $P(OMe)_3$ [168], $P(OPr)_3$ [70], $P(OPh)_3$ [70, 168], $AsPh_3$ [48], $SbPh_3$ [48], SR_2 (R = Me, Bu) [182]
$Rh_4(CO)_{11}L$	L = PPh_3 [250], P(p-Tol)$_3$ [250], $P(p\text{-}FC_6H_4)_3$ [250], $AsPh_3$ [250]
$Ir_4(CO)_{11}L$	L = PPh_3 [149]
$Co_4(CO)_{10}L_2$	L = RNC (R = Me, Et, *t*Bu, Cy, Bz) [202], PEt_3 [168], $PMePh_2$ [70], $P(OMe)_3$ [168], $P(OPr)_3$ [70], $P(OPh)_3$ [168, 224]
$Co_2Rh_2(CO)_{10}L_2$	L = PF_3 [128]
$Rh_4(CO)_{10}L_2$	L = PPh_3 [22, 250], P(p-Tol)$_3$ [250], $P(p\text{-}FC_6H_4)_3$ [250], $AsPh_3$ [250], ETPO [22], L_2 = DPE [250]
$Ir_4(CO)_{10}L_2$	L = PPh_3 [9]
$Co_4(CO)_9L_3$	L = RNC (R = Me, *t*Bu, Cy) [202], PEt_3 [168], $P(OMe)_3$ [168], $P(OPh)_3$ [70]
$Rh_4(CO)_9L_3$	L = PPh_3 [22], ETPO [22]
$Ir_4(CO)_9L_3$	L = PPh_3 [9, 177], P(*i*Pr)$_3$ [86], P(p-Tol)$_3$ [86], P(o-Tol)$_3$ [177]
$Co_4(CO)_8L_4$	L = RNC (R = Me, *t*Bu, Cy) [202], $P(OMe)_3$ [168], L_2 = ffars [76, 88], DPE[219]
$Rh_4(CO)_8L_4$	L = PPh_3 [22], ETPO [22], $P(OPh)_3$ [219], L_2 = DPE [219] L_2 = DPM[43]
$Ir_4(CO)_8L_4$	L = PEt_3 [86, 248], PBu_3 [86, 248], PPr_3 [86, 248], PMe_2Ph[46], $PMePh_2$ [46], $PEtPh_2$ [46], P(o-Tol)$_3$[219, 489], $P(OEt)_3$[219], $P(OPh)_3$[46, 219], $AsMePh_2$[46], L_2 = DPE[46, 219]
$Co_4(CO)_7L_5$	L = *t*BuNC [202]

$$Rh_4(CO)_{10}(PPh_3)_2 + 2PPh_3 + 2CO \underset{1\ atm,\ 25\ °C}{\overset{10-80\ atm,\ 50\ °C}{\rightleftharpoons}} 2Rh_2(CO)_6(PPh_3)_2 \quad (57)$$

Further substitution to give the tetra-substituted derivative is a much more delicate reaction, especially in the case of cobalt. For instance, repetition of the reported synthesis of $Co_4(CO)_8[P(OMe)_3]_4$ [168)] gave only a mixture of $Co_4(CO)_9[P(OMe)_3]_3$ and $Co_2(CO)_6[P(OMe)_3]_2$ [219)].

Table 13 shows that this behaviour is completely inverted in the case of iridium; here the monosubstituted compounds are rare and the tetrasubstituted derivatives are most commonly found. The following related reasons are believed to be responsible for this behaviour:

a) the forcing conditions necessary to obtain reaction with the sparingly soluble $Ir_4(CO)_{12}$,

b) the particular kinetic behaviour of these progressive substitutions (Section VIIIC),

c) the higher Ir-Ir bond energy.

Substitution reactions of $Ir_4(CO)_{12}$ have generally been carried out in boiling toluene and this gives the tetra-substituted derivatives[46, 86, 219)]. The di- and tri-substituted derivatives have generally been obtained by reaction of tertiary phosphines, at room temperature, with the anions $[Ir_4(CO)_{11}H]^-$ and $[Ir_8(CO)_{20}]^{2-}$ respectively[9, 86)]; reactions which probably involve the disproportionation of these anionic species. The tri-substituted derivatives have also been obtained using the reaction: –

$$2Ir_2(CO)_6L_2 \xrightarrow[\text{reflux, } N_2]{\text{Benzene}} Ir_4(CO)_9L_3 + L + 3CO \quad (L = PPh_3, P(o\text{-Tol})_3)^{156a)} \quad (58)$$

The tri-substituted derivatives, $Ir_4(CO)_9L_3$, give $Ir_4(CO)_{12}$ on reaction with CO at 550 atm and 200 °C[86, 248)], whereas the tetra-substituted derivatives, $Ir_4(CO)_8L_4$, revert only partially:

$$Ir_4(CO)_8L_4 \xrightarrow[90-100\ °C]{CO,\ 400\ atm} L + Ir_4(CO)_9L_3 \xrightarrow[200\ °C]{CO,\ 450\ atm} Ir_4(CO)_{10}L_2 + Ir_2(CO)_7L$$
$$(L = PEt_3, PBu_3)^{86,\ 248)} \quad (59)$$

The reaction between $Co_2(CO)_8$ and CS_2 is extremely complicated; one of the substances of composition $Co_4(CO)_{10}(CS_2)$ was believed to have a "butterfly" structure in which a CS_2 substitutes the C_2R_2 in $Co_4(CO)_{10}(C_2R_2)$[157, 180)]. The cluster of $Co_4(CO)_4(SEt)_8$, (Fig. 21), has been obtained from $Co(SEt)_2$ and CO (25 °C, 1 atm)[181)]. The reaction of $Co_2(CO)_8$ with bis(dimethyl-phosphine-sulphide), $(Me_2P(S)P(S)(Me_2)$, gives a green compound $Co_4(CO)_9(PMe_2)_2S$ of unknown structure[200)].

Direct substitution in $Co_4(CO)_{12}$ with arenes to give $Co_4(CO)_9$ (arene)[g] is limited by the high temperature (> 100 °C) required for the reaction and the limited thermal stability of the reaction products. The best methods seem to consist of heating (ca. 165 °C) a solution of $Co_4(CO)_{12}$ in the presence of the arene when working in a glass coil with short contact times[224]. The same synthesis is much easier when starting with $Co_2(CO)_6(PhC_2H)$ and when working at 60–70 °C in the presence of mixture of arene and norbornadiene, [for instance a 36% yield of $Co_4(CO)_9(C_6H_6)$ has been reported[151]]. At 100 °C, these compounds are demolished by CO with reformation of *some* $Co_4(CO)_{12}$ [224].

It has not been possible to obtain any evidence for the formation of the analogous compounds, $Rh_4(CO)_9$(arene)[151], probably because of their low thermal stability.

Some substituted compounds of the type $Co_4(CO)_8$(arene)$[P(OR)_3]$ (arene = = *m*-xylene, R = Me; arene = mesitylene, R = Me or Ph)[224] are known; they have been obtained using two different methods, which represent another illustration of the more severe conditions required for arene substitution:

$Co_4(CO)_9$(*m*-xylene) + $P(OMe)_3$ —(60–65 °C, heptane (yield 16%))→ $Co_4(CO)_8$(*m*-xylene)$[P(OMe)_3]$

$Co_4(CO)_{11}[P(OMe)_3]$ + *m*-xylene —(130 °C (yield 10%))→ $Co_4(CO)_8$(*m*-xylene)$[P(OMe)_3]$ (60)

The compound $Rh_4(CO)_8$(COT)[156] has also been reported.

The reaction of $Co_4(CO)_{12}$ with acetylenes[84] readily gives compounds of the type, $Co_4(CO)_{10}(C_2R'R'')$ (R' = R'' = H[167], Me[83], CD_3 [215], CF_3 [83], Et[167], Ph[167]; R' = H, R'' = $SiMe_3$ [167], Me[82], CF_3 [82], Ph[82], C_6F_5 [82]; R' = Me, R'' = CF_3 [83]; R' = Ph, R'' = COOMe[167]; $C_2R'R''$ = C_6F_4 [216]) when using the stoichiometric amount of the acetylene and when working at 20–100 °C[167]. An excess of the acetylene gives $Co_2(CO)_6(C_2R'R'')$ derivatives with contemporary catalytic trimerisation of the acetylene[167]. The analogous rhodium derivatives, $Rh_4(CO)_{10}(C_2R'R'')$ (R' = R'' = CF_3, Ph)[22], are known and, in the trimerisation of acetylenes, the catalytic activity of $Rh_4(CO)_{12}$ is superior to that of $Co_4(CO)_{12}$ but the selectivity is lower[135].

---

[g] Arene = benzene[20, 151, 224], toluene[23, 151], *p*-xylene[151, 224], *m*-xylene[151, 224], mesitylene[23, 151], durene[151], *iso*-durene[224], pentamethylbenzene[151], hexamethylbenzene[151], ethylbenzene[151], biphenyl[224], tetrahydronophthalene[23], phenanthrene[224], fluorene[224], anisole[151, 224], *o*-Me-C_6H_4OMe[151], *p*-Me-C_6H_4OMe[151], 3,5-Me_2-C_6H_3-OMe[151], cycloheptatriene[156].

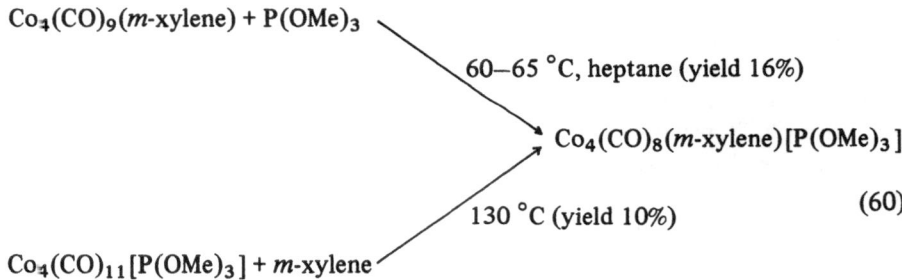

There has also been a report of the following interesting reaction which involves the dimerisation of the carbyne ligand[215]:

$$2Co_3(CO)_6(arene)(\mu_3\text{-}CCD_3) \xrightarrow{110\,°C} Co_4(CO)_{10}[C_2(CD_3)_2] + 2Co + 2CO + 2\,arene \qquad (61)$$

$Rh_4(CO)_{12}$ reacts with α-olefins in the presence of hydrogen according to the following scheme:

$$3Rh_4(CO)_{12} + 4R\text{–}CH\text{=}CH_2 + 4H_2 \xrightarrow[25\,°C]{hexane} 2R\text{–}CH_2\text{–}CH_2\text{–}CHO + 2R\text{–}CH(CH_3)\text{–}CHO + 2Rh_6(CO)_{16} \qquad (62)$$

This reaction is first order both in $Rh_4(CO)_{12}$ and hydrogen and is inhibited by CO[58, 77]; using the same conditions $Co_4(CO)_{12}$ is inactive.

The cluster $Co_4(CO)_2Cp_4$ has been obtained according to Eq. (6). The molecular weight has been confirmed by mass spectrometry, and infrared spectra are consistent with the carbonyls being triply bridging[237, 238].

E. Heterometallic Homologous Clusters

There are several examples of mixed clusters which are homologous with the dodecacarbonyls. Some of these compounds, such as the anion $[FeCo_3(CO)_{12}]^-$ and the neutral cyclopentadienyl carbonyl $Co_3Ni(CO)_9Cp$, are believed to be *iso*-structural with $Co_4(CO)_{12}$ and it is therefore convenient to discuss these compounds now.

The first mixed carbonyl, the anion $[FeCo_3(CO)_{12}]^-$, was obtained using the redox condensation shown in Eq. (63)[52].

$$7/2Co_2(CO)_8 + 2Fe(CO)_5 \xrightarrow[reflux]{acetone} [Co(acetone)_6][FeCo_3(CO)_{12}]_2 + 14CO \quad \text{(yield ca. 90\%)} \qquad (63)$$

The stoichiometry of this reaction has been carefully studied[52] but the mechanism is obscure. The more probable hypothesis is:

$$[Co_3(CO)_{10}]^- + Fe(CO)_5 \longrightarrow [FeCo_3(CO)_{12}]^- + 3CO \qquad (64)$$

Addition of strong acids to an aqueous solution of this anion results in the nearly quantitative separation of the insoluble hydride, $FeCo_3(CO)_{12}H$;[52] the corresponding deuteride has also been prepared[160, 189].

The preparation of analogous compounds containing ruthenium and osmium has been less studied: $RuCo_3(CO)_{12}H$ has been obtained in 7% yield both from $Co_2(CO)_8$ and $Ru_3(CO)_{12}$ in refluxing acetone followed by acidification[189], and from

$Co_2(CO)_8$ and $[Ru(CO)_3Cl_2]_2$ in refluxing THF[252]. Reduction of $Na_2[OsCl_6]$ with $Co_2(CO)_8$ and CO, followed by acidification, gives the analogous compound $OsCo_3(CO)_{12}H$[160].

The hydride $FeCo_3(CO)_{12}H$ (decomp. ca. 80°) behaves as a strong acid, and for a cobalt containing cluster is remarkably stable to air (possibly for steric reasons)[52]. It reacts with tertiary phosphines and phosphites, best in refluxing chloroform, to give substitution products $FeCo_3(CO)_{12-n}L_nH$, (n = 1 and 2, L = PEt_3, $PMePh_2$, PPh_3, $P(OMe)_3$, $P(OPr)_3$, $P(OPh)_3$; n = 3, L = $PMePh_2$)[70, 132]. Mössbauer spectra (Section VB) indicate that, in all cases, substitution takes place only in the basal plane containing the cobalt atoms, (Fig. 13)[70]. Using DPE the tetra-substituted product $FeCo_3(CO)_8(DPE)_2H$ has been obtained[70]. The high stability of this cluster is confirmed by the direct substitution of carbon monoxide in the corresponding anion:

$$[NEt_4][FeCo_3(CO)_{12}] \xrightarrow[\text{acetone, reflux}]{L} [NEt_4][FeCo_3(CO)_{11}L] + CO \qquad (65)$$

(L = $PMePh_2$, PPh_3, $P(OPr)_3$, yield 60–95%)

The anion $[FeCo_3(CO)_{10}L_2]^-$ (L_2 = DPE and C_2Ph_2) has also been obtained similarly.

The reaction between $Co_2(CO)_8$ and $Ni_2(CO)_2Cp_2$ in boiling petroleum ether gives $Co_3Ni(CO)_9Cp$ (yield ca. 20%)[129]. The molecular weight of this dark green compound has been confirmed by mass spectrometry and the infrared spectrum shows bridging carbonyl absorptions in the same region as found in $Co_4(CO)_{12}$, (1850 cm^{-1}). The most probable structure for $Co_3Ni(CO)_9Cp$ involves replacement of the apical $Co(CO)_3$ in $Co_4(CO)_{12}$ by the NiCp group[129]. Some substituted derivatives, $Co_3Ni(CO)_8(RNC)Cp$ (R = Me, *t*-Bu)[202], are also known.

The redox condensation shown in Eq. (9) gives rise to the dianion $[Fe_3Ni(CO)_{12}]^{2-}$ and it has also been possible to isolate salts of the corresponding hydride, $[Fe_3Ni(CO)_{12}H]^-$ (30.5 τ). The presence of infrared absorptions due to bridging carbonyls and the complexity of the infrared spectrum indicate a structure similar to $Co_4(CO)_{12}$ with the nickel in the basal plane[171].

XIII. Clusters of Ni, Pd, Pt and Mixed Clusters Containing Group VIII C Metals

In this triad the simple polynuclear neutral carbonyls are unknown: the compound $[Pt(CO)_2]_n$ probably corresponds to a mixture of anions $[Pt_3(CO)_6]_n^{2-}$ (n ≃ 10)[60]. Stable polynuclear species have been obtained only in the presence of ligands less π-acidic than CO, or as anions.

Although the synthesis of the anion $[Ni_4(CO)_9]^{2-}$ [for instance by reduction of $Ni(CO)_4$ with sodium amalgam in THF] was reported in 1960[124, 126] a later study showed this supposed red tetranickel cluster to be the hexanuclear species $[Ni_6(CO)_{12}]^{2-}$[38]. This does not mean tetranuclear carbonyl nickelate anions do not exist, but only that such species have not as yet been isolated and characterized.

Spontaneous decomposition of $Ni(CO)_2(C_4F_6)$ gives the tetranuclear derivative, $Ni_4(CO)_4(C_4F_6)_3$ (C_4F_6 = hexafluoro-but-2-ene)[79, 153]. It is a sublimable purple-red substance which has the structure shown in Fig. 16.

The cluster, $Ni_4(CO)_6[P(CH_2CH_2CN)_3]_4$, has been obtained in low yield (ca. 5%) by the reaction of $Ni(CO)_4$ with $P(CH_2CH_2CN)_3$ in boiling methanol[17, 183], and the structure of this compound has already been reported in Fig. 12. Whereas the nickel species, $Ni(CO)_2(PR_3)_2$, are generally stable, the corresponding platinum species decompose spontaneously[50]:

$$5Pt(CO)_2(PR_3)_2 \underset{CO}{\overset{N_2}{\rightleftarrows}} Pt_3(CO)_3(PR_3)_4 + 2Pt(PR_3)_3 + 7CO \qquad (66)$$

A further condensation occurs on reaction of the pure trinuclear compounds with CO:

$$2Pt_3(CO)_3(PR_3)_4 + 3CO \xrightarrow[1\ atm]{25\ °C} Pt_4(CO)_5(PR_3)_4 + 2Pt(CO)_2(PR_3)_2 \qquad (67)$$
$$(PR_3 = PMe_2Ph, PPh_3)$$ [50]

There are at least two different series of isomeric compounds having the formula $Pt_4(CO)_5L_4$. One series, which is brown-black (L = PMe_2Ph [50], PPh_3 [50, 251], $PPh_2(o\text{-}Tol)$[251], $PPh(o\text{-}Tol)_2$ [251]) and has only bridging carbonyl infrared absorptions, has the structure shown in Fig. 19. The other series of compounds, which are all red, (L = PEt_3, PPh_2Bz, $AsPh_3$)[50] have infrared absorptions due to the presence of both terminal and bridging carbonyls[50]. Other compounds formulated as $Pt_4(CO)_5(AsPh_3)_3$ [50], $Pt_4(CO)_6(PPh_3)_3$ [35], and $Pt_4(CO)_8(PPh_3)_3$ [35] have not yet been well characterized.

In addition to the mixed compounds, $[Fe_3Ni(CO)_{12}]^{2-}$ and $Co_3Ni(CO)_9Cp$, mentioned in the preceding section, several other mixed tetranuclear compounds containing metals of this triad exist. The reaction of an ethanolic solution of $NiCl_2$ with $[Co(CO)_4]^-$ gives a brown solution containing the anion $[Co_3Ni(CO)_{11}]^-$, which has been isolated as the tetramethylammonium salt[59]. The infrared spectrum shows the presence of terminal, edge- and face-bridging carbonyls, but the structure is presently unknown. This anion is readily demolished by CO, and its more remarkable reaction is the equilibrium:[59]

$$2[Co_3Ni(CO)_{11}]^- \rightleftarrows [Co_4Ni_2(CO)_{14}]^{2-} + Co_2(CO)_8 \qquad (68)$$

The reaction between $Ni_2Cp_2(C_2R_2)$ and $Fe_3(CO)_{12}$ in refluxing benzene produces a series of black compounds formulated as $Fe_2Ni_2(CO)_6Cp_2(C_2R_2)$ (C_2R_2 = C_2HPh, C_2Ph_2, $[C_2Ph]_2$)[231].

The infrared spectra of these compounds show that there are no bridging carbonyl groups present and the structure shown in Fig. 24, which is similar to that of $Co_4(CO)_{10}(C_2R_2)$, has been proposed.

The cluster $Os_2Pt_2(CO)_8(PPh_3)_2H_2$ has been obtained in 8% yield starting from $Os(CO)_4H_2$ and $Pt(PPh_3)_2(C_2H_4)$ in benzene at 25 °C. It is a yellow-orange solid

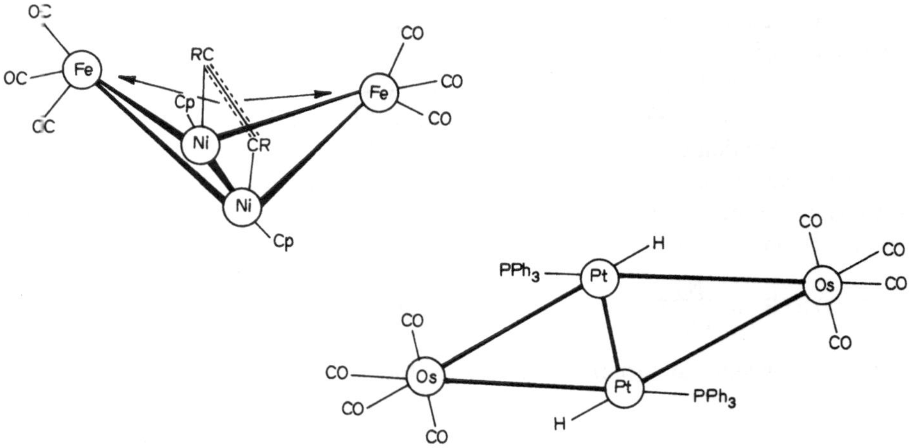

Fig. 24. Proposed structures for $Fe_2Ni_2(CO)_6Cp_2(C_2R_2)$[231] and $Os_2Pt_2(CO)_8(PPh_3)_2H_2$[34]

(17.76 τ), with no infrared absorptions due to bridging carbonyls[34], and a "butterfly" structure seems possible (Fig. 24).

The cluster $Co_2Pt_2(CO)_8(PPh_3)_2$ shown in Fig. 19 has been prepared from *trans*-$PtCl_2(PPh_3)_2$ and $Na[Co(CO)_4]$ in THF[30]. Similarly the reaction of *trans*-MCl_2-$(PPh_3)_2$ (M = Pt[30], Pd[81]) and Na $[Mo(CO)_3Cp]$ gives rise to substances formulated as $Mo_2M_2(CO)_6Cp_2(PPh_3)_2$ but which are not yet completely characterized.

XIV. References

[1] Abel, E. W., Hender, P. J., McLean, R. A. M., Quarashi, M. M.: Inorganica Chimica Acta *3*, 77 (1969)
[2] Aime, S., Gambino, O., Milone, L., Sappa, E., Rosenberg E.: Inorganica Chimica Acta *15*, 53 (1975)
[3] Albano, V. G., Bellon, P. L., Scatturin, V.: Chem. Comm. **1967**, 730
[4] Albano, V. G., Ciani, G., Martinengo, S.: J. Organometall. Chem. *78*, 265 (1974)
[5] Albano, V. G., Ciani, G., Freni, M. Romiti, P.: J. Organometall. Chem. *96*, 259 (1975)
[6] Albano, V. G., Ciani, G., Anker, P., Martinengo, S., Fumagalli, A.: J. Organometall. Chem. *116*, 343 (1976)
[7] Alich, A., Nelson, N. J., Strape, D. Shriver, D. F.: Inorg. Chem. *11*, 2976 (1972)
[8] Anders, U., Graham, W. A. G.: Chem. Comm. **1966**, 291
[9] Angoletta, M., Malatesta, L., Caglio, G.: J. Organometall. Chem. *94*, 99 (1975)
[10] Bätzel, V., Müller, U., Allmann, R.: J. Organometall. Chem. *102*, 109 (1975)
[11] Bau, R., Fontal, B., Kaesz, H. D., Churchill, M. R.: J. Amer. Chem. Soc. *89*, 6374 (1967)
[12] Bau, R., Kirtley, S. W., Sorrell, T. N., Winarko, S.: J. Amer. Chem. Soc. *96*, 988 (1974)
[13] Behrens, H., Weber, R.: Z. Anorg. Allg. Chem. *281*, 190 (1955)
[14] Behrens, H., Haag, W.: Chem. Ber. *94*, 312 (1961)
[15] Behrens, H., Wakamatsu, H.: Ber. *99*, 2753 (1966)
[16] Belford, R., Taylor, H. P., Woodward, P.: J. Chem. Soc. Dalton **1972**, 2425
[17] Bennett, M. J., Cotton, F. A., Winquist, B. H.: J. Amer. Chem. Soc. *89*, 5366 (1967)
[18] Blackmore, T., Cotton, J. D., Bruce, M. I., Stone, F. G. A.: J. Chem. Soc. (A) **1968** 2931
[18a] Blount, J. F., Dahl, L. F., Hoogzand, C., Hubel, W.: J. Amer. Chem. Soc. *88*, 292 (1966)
[19] Bigorgne, M.: J. Organometall. Chem. *94*, 161 (1975)
[20] Bird, B. H., Fraser, A. R.: J. Organometall. Chem. *73*, 103 (1974)
[21] Booth, B. L., Else, M. J., Fields, R., Goldwhite, H., Haszeldine, R. N.: J. Organometall. Chem. *14*, 417 (1968)
[22] Booth, B. L., Else, M. J., Fields, R., Haszeldine, R. N.: J. Organometall. Chem. *87*, 119 (1971)
[23] Bor, G., Sbrignadello, G., Mercati, F., J. Organometall. Chem. *46*, 357 (1972)
[24] Bor, G., Dietler, V. K., Pino, P.: Abstracts of VII Internat. Conf. Organometall. Chem. Venice 1975, p. 27
[25] Bor, G., Sbrignadello, G., Noak, K.: Helv. Chim. Acta. *58*, 815 (1975)
[26] Bradford, C. W., von Bronswijk, W., Clark, R. J. H., Nyholm, R. S.: J. Chem. Soc. (A) **1970**, 2889
[27] Bradford, C. W., Nyholm, R. S.: J. Chem. Soc. (A) **1971**, 2038
[28] Braddock, J. N., Meyer, T. J.: Inorg. Chem. *12*, 723 (1973)
[29] Braterman, P. S.: Metal carbonyl spectra. New York: Academic Press 1975
[30] Braunstein, P., Dehand, J., Hennig, J. F.: J. Organometall. Chem. *92*, 117 (1975)
[31] Bright, D.: Chem. Comm. **1970**, 1169
[32] Bruce, M. I.: Adv. Organometall. Chem. *6*, 273 (1968)
[33] Bruce, M. I., Cairns, M. A., Green, M.: J. Chem. Soc. Dalton **1972**, 1293
[34] Bruce, M. I., Shaw, G., Stone, F. G. A.: J. Chem. Soc. Dalton **1972**, 1781
[35] Bruce, M. I., Shaw, G., Stone, F. G. A.: J. Chem. Soc. Dalton **1972**, 1082
[36] Bruce, M. I., Shaw, G., Stone, F. G. A.: J. Chem. Soc. Dalton **1973**, 1667
[37] Bullitt, J. G., Cotton, F. A., Marks, T. J.: Inorg. Chem. *11*, 671 (1972)
[38] Calabrese, J., Cavalieri, A., Chini, P., Dahl, L. F., Longoni, G., Martinengo, S.: J. Amer. Chem. Soc. *96*, 2516 (1974)
[39] Calderazzo, F., Ercoli, R., Natta, G.: Organic syntheses via metal carbonyls. Vol.1. Wender, I. and Pino, P. (eds.). New York: Interscience-Wiley 1968, pp. 1–272
[40] Canty, A. J., Domingos, A. J. P., Johnson, B. F. G., Lewis, J.: J. Chem. Soc. Dalton **1973**, 2056
[41] Canty, A. J., Johnson, B. F. G., Lewis, J., Norton, J. R.: Chem. Comm. **1972**, 1331

42) Cariati, F., Fantucci, P., Vananzi, V., Barone, P.: Rend. Ist. Lomb. Sci (A) *105*, 122 (1971)
43) Carré, F. H., Cotton, F. A., Frenz, B. A.: Inorg. Chem. *15*, 380 (1976)
44) Cartner, A., Cunningham, R. G., Robinson, B. H.: J. Organometall. Chem. *92*, 49 (1975)
45) Cattermole, P. E., Osborne, A. G.: J. Organometall. Chem. *37*, C17 (1972)
46) Cattermole, P. E., Orrell, K. G., Osborne, A. G.: J. Chem. Soc. Dalton **1974**, 328
47) Cetini, G., Gambino, O., Sappa, E., Vaglio, G.: Ricerca Sci. *37*, 430 (1967)
48) Cetini, G., Gambino, O., Rossetti, R., Stanghellini, P. L.: Inorg. Chem. *7*, 609 (1968)
49) Chaston, S. H., Stone, F. G. A.: J. Chem. Soc. (A) **1969**, 500
50) Chatt, J., Chini, P.: J. Chem. Soc. (A) **1970**, 1538
51) Chini, P.: Abstracts XVII I.U.P.A.C. Congress. Vol. 1. Weinheim Verlag Chemie 1959, p. 23
52) Chini, P., Colli, L., Paraldo, M.: Gazz. chim. Ital. *90*, 1005 (1960)
53) Chini, P., Albano, V.: J. Organometall. Chem. *15*, 433 (1968)
54) Chini, P.: Inorganica Chimica Acta Rev. *2*, 31 (1968) and references therein
55) Chini, P., Albano, V. G., Martinengo, S.: J. Organometall. Chem. *16*, 471 (1969)
56) Chini, P., Martinengo, S.: Inorganica Chimica Acta *3*, 71 (1969)
57) Chini, P., Martinengo, S.: Inorganica Chimica Acta *3*, 315 (1969)
58) Chini, P., Martinengo, S., Garlaschelli, G.: Chem. Comm. **1972**, 709
59) Chini, P., Cavalieri, A., Martinengo, S.: Coord. Chem. Rev. *8*, 3 (1972)
60) Chini, P., Longoni, G., Albano, V. G.: Adv. Organometall. Chem. **XIV**, 285 (1976)
61) Chini, P., Ceriotti, A., Martinengo, S.: unpublished observations
62) Churchill, M. R., Bird, P. H.: J. Amer. Chem. Soc. *80*, 800 (1968)
63) Churchill, M. R., Gold, K., Bird, P. H.: Inorg. Chem. *80*, 1956 (1969)
64) Churchill, M. R., Veidis, M. V.: Chem. Comm. **1970**, 1470
65) Churchill, M. R., Veidis, M. V.: J. Chem. Soc. (A) **1971**, 2170
66) Cohen, M., Kidd, D. R., Brown, T. L.: J. Amer. Chem. Soc. *97*, 4408 (1975)
67) Colton, R., Commons, C. J., Hoskins, B. F.: J. Chem. Soc. Chem. Comm. **1975**, 363
68) Connor, J. A., Skinner, H. A., Virmani, Y.: Faraday Symp. Chem. Soc. *8*, 18 (1974)
69) Cooke, C. G., Mays, M. J.: J. Organometall. Chem. *74*, 449 (1974)
70) Cooke, C. G., Mays, M. J.: J. Chem. Soc. Dalton **1975**, 455
71) Corradini, P.: J. Chem. Phys. *31*, 1676 (1959)
72) Cotton, F. A., Monchamp, R. R.: J. Chem. Soc. **1960**, 1882
73) Cotton, F. A., Kruczynski, L., Shapiro, B. L., Johnson, L. F.: J. Amer. Chem. Soc. *94*, 6191 (1972)
74) Cotton, F. A., Troup, J. M.: J. Amer. Chem. Soc. *96*, 1233 (1974)
75) Cotton, F. A., Hunter, D. L., White, A. J.: Inorg. Chem. *14*, 703 (1975)
76) Crow, J. P., Cullen, W. R.: Inorg. Chem. *10*, 2165 (1971)
77) Csontos, G., Heil, B., Markó, L., Chini, P.: Hung. J. Ind. Chem. (Veszprem) *1*, 53 (1973)
78) Dahl, L. F., Smith, D. L.: J. Amer. Chem. Soc. *84*, 2450 (1962)
79) Davidson, J. L., Green, M., Stone, F. G. A., Welch, A. J.: J. Amer. Chem. Soc. *97*, 7490 (1975)
80) Davis, R.: J. Organometall. Chem. *78*, 237 (1974) and references therein
81) Dehand, J., Pfeiffer, M.: J. Organometall. Chem. *104*, 377 (1976)
82) Dickson, R. S., Tailby, G. R.: Aust. J. Chem. *23*, 229 (1970)
83) Dickson, R. S., Fraser, P. J.: Aust. J. Chem. *23*, 2403 (1970)
84) Dickson, R. S., Fraser, P. J.: Adv. Organometall. Chem. *12*, 322 (1974)
85) Doedens, J., Dahl, L. F.: J. Amer. Chem. Soc. *88*, 4847 (1966)
86) Drakesmith, A. J., Whyman, R.: J. Chem. Soc. Dalton **1973**, 362
87) Eady, C. R., Johnson, B. F. G., Lewis, J.: J. Organometall. Chem. *57*, C84 (1973)
88) Einstein, F. W. B., Jones R. D., J. Chem. Soc. (A) **1971**, 3359
89) Ercoli, R., Barbieri-Hermitte, F.: Rend. Acad. Lincei Clas. Sci. Fis. Mat. Nat. *16*, 249 (1954)
90) Ercoli, R., Chini, P., Massimauri, M., Chim. Ind. Milano *41*, 132 (1959)
91) Evans, J., Johnson, B. F. G., Lewis, J., Norton, J. R., Cotton, F. A.: J.C.S. Chem. Comm. **1973**, 807
92) Evans, J., Johnson, B. F. G., Lewis, J., Matheson, T. W.: J. Amer. Chem. Soc. *97*, 1245 (1975)

93) Farmery, K., Kilner, M., Greatrex, R., Greenwood, N. N.: J. Chem. Soc. (A) **1969,** 2339
94) Ferguson, J. A., Meyer, T. J.: J. Amer. Chem. Soc. *94,* 3409 (1972)
95) Fischer, J., Mitschler, A., Weiss, R., Dehand, J., Nannig, J. F.: J. Organometall. Chem. *91,* C37 (1975)
96) Fontal, A.: Dissertation, University of California, Los Angeles 1969
97) Gansow, O. A., Burke, A. R., Vernon, W. D.: J. Amer. Chem. Soc. *94,* 2250 (1972)
98) Gardner, P. J., Cartner, A., Cunninghame, R., Robinson, B. H.: J. Chem. Soc. Dalton **1975,** 2582
99) Gee, J., Powell, H. M.: quoted as a personal communication in Ref.[26]
100) Gilmore, C. J., Woodward, P.: J. Chem. Soc. (A) **1971,** 3453
101) Ginsberg, A. P., Hawkes, M. K.: J. Amer. Chem. Soc. *90,* 5930 (1968)
102) Giordano, G., Canziani, F., Martinengo, S., Albano, V. G., Ciani, G., Manassero, M.: unpublished work
103) Greatrex, R., Greenwood, N. N.: Discuss. Farad. Soc. *47,* 126 (1969)
104) Haas, H., Sheline, R. K.: J. Inorg. Nucl. Chem. *29,* 693 (1967)
105) Haines, L. M., Stiddard, M. H. B.: Adv. Inorg. and Radiochem. *12,* 53 (1969)
106) Harrison, W., Marsh, W. C., Trotter, J.: J. Chem. Soc. Dalton **1972,** 1009
107) Hanlan, L. A., Ozin, G. A.: J. Amer. Chem. Soc. *96,* 6324 (1974)
108) Haupt, H. J., Neumann, F., Preut, H.: J. Organometall. Chem. *99,* 439 (1975)
109) Hayter, R. G.: J. Amer. Chem. Soc. *88,* 4376 (1966)
110) Heaton, B. T., Towl, A. D. C., Chini, P., Fumagalli, A., McCaffrey, D. J. A., Martinengo, S.: J.C.S. Chem. Comm. **1975,** 523
111) Heaton, B. T., Longoni, G., Martinengo, S., Fumagalli, A., Chini, P.: unpublished results
112) Hein, F., Reinart, H.: Ber. *93,* 2089 (1960)
113) Herberhold, M., Süss, G.: Angew. Chem. Internat. Ed. *14,* 700 (1975)
114) Hieber, W., Lagally, H.: Z. Anorg. Allg. Chem. *245,* 321 (1940)
115) Hieber, W., Lagally, H.: Z. Anorg. Allg. Chem. *251,* 96 (1943)
116) Hieber, W., Sedlemeier, J.: Ber. *87,* 25 (1954)
117) Hieber, W., Schuster, L.: Z. Anorg. Allg. Chem. *285,* 205 (1956)
118) Hieber, W., Brendel, G.: Z. Anorg. Allg. Chem. *289,* 324 (1957)
119) Hieber, W., Werner, R.: Ber. *90,* 286 (1957)
120) Hieber, W., Floss, J. G.: Ber. *90,* 1617 (1957)
121) Hieber, W., Kahlen, N.: Ber. *91,* 2223 (1958)
122) Hieber, W., Lipp, A.: Ber. *92,* 2075 (1959)
123) Hieber, W., Lipp, A.: Ber. *92,* 2085 (1959)
124) Hieber, W., Kroder, W., Zahn, E.: Z. Naturforsch. *15b,* 325 (1960)
125) Hieber, W., Kruck, T.: Chem. Ber. *95,* 2027 (1962)
126) Hieber, W., Eldermann, J., Zahn, E.: Z. Naturforsch. *18b,* 589 (1963)
127) Hieber, W., Shubert, H.: Z. Anorg. Allg. Chem. *338,* 37 (1965)
128) Hosseini, H. E., Nixon, J. F.: J. Organometall. Chem. *97,* C24 (1975)
129) Hsieh, A. T. T., Knight, J.: J. Organometall. Chem. *26,* 125 (1971)
130) Hsieh, A. T. T., Mays, M. J.: J. Organometall. Chem. *39,* 157 (1972)
131) Hsieh, A. T. T.: Inorganica Chimica Acta *14,* 87 (1975)
132) Huie, B. T., Knobler, C. B., Kaesz, H. D.: J.C.S. Chem. Comm. **1975,** 684
133) Huttner, G., Lorenz, H.: Chem. Ber. *107,* 996 (1974)
134) Huttner, G., Lorenz, H.: Chem. Ber. *108,* 973 (1975)
135) Iwashita, Y., Tamura, F.: Bull. Chem. Soc. Japan *43,* 1517 (1970)
136) Jackson, W. G., Johnson, B. F. G., Kelland, J. W., Lewis, J., Schorpp, K. T.: J. Organometall. Chem. *88,* C17 (1975)
137) Johnson, B. F. G., Johnston, R. D., Lewis, J.: J. Chem. Soc. (A) **1968,** 2865
138) Johnson, B. F. G., Johnston, R. D., Lewis, J., Robinson, B. H., Wilkinson, G.: J. Chem. Soc. (A) **1968,** 2856
139) Johnson, B. F. G., Lewis, J., Kilty, P. A.: J. Chem. Soc. (A) **1968,** 2859
140) Johnson, B. F. G., Lewis, J., Williams, I. G.: J. Chem. Soc. (A) **1970,** 901
141) Johnson, B. F. G., Lewis, J., Matheson, T. W.: J.C.S. Chem. Comm. **1974,** 441
142) Johnson, B. F. G., Lewis, J., Schorpp, K. T.: J. Organometall. Chem. *91,* C13 (1975)

143) Johnson, B. F. G.: Mass spectrometry of metal compounds, Charalambous, J. (Ed.). London: Butterworths 1975, pp. 113–125
144) Johnson, B. F. G.: J. C. S. Chem. Comm. **1976,** 211
145) Kaesz, H. D., Fontal, B., Bau, R., Kirtley, S. W., Churchill, M. R.: J. Amer. Chem. Soc. *91,* 1021 (1969)
146) Kaesz, H. D., Knox, S. A. R., Keopke, J. W., Saillant, R. B.: Chem. Comm. **1971,** 477
147) Kaesz, H. D., Saillant, R. B.: Chem. Rev. *72,* 231 (1972)
148) Kaesz, H. D.: Chem. in Brit. *9,* 344 (1973) and references therein
149) Karel, K., Norton, G.: J. Amer. Chem. Soc. *96,* 6812 (1974)
150) Kaska, W. C.: J. Amer. Chem. Soc. *90,* 6340 (1968)
151) Khand, I. U., Knox, G. R., Pauson, P. L., Watts, W. E.: J. Chem. Soc. Perkin I **1973,** 975
152) Kiefer, W., Bernstein, H. J.: Appl. Spect. *25,* 500 and 609 (1971)
153) King, R. B., Bruce, M. I., Phillips, J. R., Stone, F. G. A.: Inorg. Chem. *5,* 684 (1966)
154) King, R. B., Inorg. Chem. *5,* 2227 (1966)
155) King, R. B., Kapoor, P. N.: Inorg. Chem. *11,* 337 (1972)
156) Kitamura, T., Joh, J.: J. Organometall. Chem. *65,* 235 (1974)
157) Klumpp, E., Bor, G., Markó, L.: J. Organometall. Chem. *11,* 207 (1968)
158) Knight, J., Mays, M. J.: Chem. and Ind. *9,* 115 (1968)
159) Knight, J., Mays, M. J.: J. Chem. Soc. (A) **1970,** 654
160) Knight, J., Mays, M. J.: J. Chem. Soc. (A) **1970,** 711
161) Knight, J., Mays, M. J.: J. Chem. Soc. Dalton **1972,** 1022
162) Knox, S. A. R., Kaesz, H. D.: J. Amer. Chem. Soc. *93,* 4594 (1971)
163) Knox, S. A. R., Koepke, J. W., Andrews, M. A., Kaesz, H. D.: J. Amer. Chem. Soc. *97,* 3942 (1975)
164) Koepke, J. W., Johnson, J. R., Knox, S. A. R., Kaesz, H. D.: J. Amer. Chem. Soc. *97,* 3947 (1975)
165) Kristoff, J. S., Hirsekorn, F. J., Muetterties, E. L.: J. Amer. Chem. Soc. *97,* 2571 (1975)
166) Kristoff, J. S., Shriver, D. F.: Inorg. Chem. *13,* 499 (1974)
167) Kruerke, V., Hubel, W.: Ber. *94,* 2829 (1961)
168) Labrone, D., Poilblanc, R.: Inorganica Chimica Acta *6,* 387 (1972)
169) Lindner, E., Behrens, H., Birkle, S.: J. Organometall. Chem. *15,* 165 (1968)
170) Litzow, M. R., Spalding, T. R.: Mass Spectrometry of Inorganic and organometallic compounds. Lappert, M. F. (ed.). Amsterdam: Elsevier 1973
171) Longoni, G., Chini, P.: in press
172) Longoni, G., Chini, P.: unpublished results
173) Lucken, E. A. C., Noack, K., Williams, D. F.: J. Chem. Soc. (A) **1967,** 148
174) Malatesta, L., Caglio, G.: Chem. Comm. **1967,** 420
175) Malatesta, L., Caglio, G., Angoletta, M.: Chem. Comm. **1970,** 532
176) Malatesta, L., Caglio, G., Angoletta, M.: Inorg. Synth. *13,* 95 (1972)
177) Malatesta, L., Angoletta, M., Caglio, G.: J. Organometall. Chem. *73,* 265 (1974)
178) Manassero, M., Longoni, G., Sansoni, M.: J. C. S. Chem. Comm. *1976*, 919
179) Markó, L., Bor, G., Almasy, G.: Ber. *94,* 847 (1961)
180) Markó, L., Bor, G., Klummp, E.: Angew. Chem. *75,* 248 (1963)
181) Markó, L., Bor, G.: J. Organometall. Chem. *3,* 162 (1965)
182) Markó, L.: Acta Chim. Acad. Sci. Hung. *59,* 389 (1969)
183) Martinengo, S., Chini, P., Giordano, G.: J. Organometall. Chem. *27,* 389 (1971)
184) Martinengo, S., Chini, P., Albano, V. G., Cariati, F., Salvatori, T.: J. Organometall. Chem. *59,* 379 (1973)
185) Martinengo, S., Fumagalli, A., Chini, P., Albano, V. G., Ciani, G.: J. Organometall. Chem.: *116*, 133 (1976)
186) Mason, R., Thomas, K. M.: J. Organometall. Chem. *43,* C39 (1972)
187) Mason, R., Mingos, D. M. P.: J. Organometall. Chem. *50,* 53 (1973)
188) Matheson, T. W., Robinson, B. H.: J. Organometall. Chem. *88,* 367 (1975)
189) Mays, M. J., Simpson, R. N. F.: J. Chem. Soc. (A) **1968,** 1444
190) McCleverty, J. A., Wilkinson, G.: Inorg. Synth. *8,* 211 (1966)

191) Meriwether, L. S., Colthup, E. C., Fiene, M. L., Cotton, F. A.: J. Inorg. Nucl. Chem. *11*, 181 (1959)
192) Meyer, T. J.: Prog. Inorg. Chem. **XIX**, 1 (1975)
193) Milone, L., Aime, S., Randall, E. W., Rosenberg, E.: J.C.S. Chem. Comm. **1975**, 452
194) Moelwyn-Hughes, T., Garner, A. W. B., Gordon, N.: J. Organometall. Chem. *26*, 373 (1971)
195) Mond, L., Hirtz, H., Cowap, M. D.: J. Chem. Soc. *92*, 798 (1910)
196) Moss, J. R., Graham, W. A. G.: J. Organometall. Chem. *23*, C47 (1969)
197) Moss, J. R., Graham, W. A. G.: J. Organometall. Chem. *23*, C23 (1970)
198) Müller, J., Dorner, H., Huttner, G., Lorenz, H.: Angew. Chem. Internat. Ed. *12*, 1005 (1973)
199) Müller, J., Dorner, H.: Angew. Chem. Internat. Ed. *12*, 843 (1973)
200) Natile, G., Pignataro, S., Innorta, G., Bor, G.: J. Organometall. Chem. *40*, 215 (1972)
201) Neuman, M. A., Trinh-Toan, Dahl, L. F.: J. Amer. Chem. Soc. *94*, 3383 (1972)
202) Newman, J., Manning, A. R.: J. Chem. Soc. Dalton **1974**, 2549
203) Noak, K.: J. Organometall. Chem. *7*, 151 (1967)
204) Norton, J. R., Collman, J. P., Dolcetti, G., Robinson, W. T.: Inorg. Chem. *11*, 382 (1972)
205) Norton, J. R., Collman, J. P.: Inorg. Chem. *12*, 476 (1973)
205a) Onaka, S., Shriver, D. F.: Inorg. Chem. *15*, 915 (1976)
206) Osborne, A. G., Stiddard, M. B. H.: J. Organometall. Chem. *3*, 340 (1965)
207) Palyi, G., Piacenti, F., Markó, L.: Inorganica Chimica Acta Rev. *4*, 109 (1970)
208) Peake, B. M., Robinson, B. H., Simpson, J., Watson, D.: J.C.S. Chem. Comm. **1974**, 945
209) Penfold, R., Robinson, B. H.: Acc. Chem. Res. *6*, 73 (1973)
210) Picacenti, F., Bianchi, M., Frediani, P., Benedetti, F.: Inorg. Chem. *10*, 2759 (1971)
211) Potenza, J., Giordano, P., Mastropaolo, D., Efraty, A., King, R. B.: J.C.S. Chem. Comm. **1972**, 1333
212) Preut, H., Haupt, H. J.: Chem. Ber. *107*, 2860 (1974)
213) Quicksall, C. O., Spiro, T. G.: Inorg. Chem. *7*, 2365 (1968)
214) Quicksall, C. O., Spiro, T. G.: Inorg. Chem. *8*, 2011 (1969)
215) Robinson, B. H., Spencer, J. L.: J. Chem. Soc. (A) **1971**, 2045
216) Rue, D. H., Messey, A. G.: J. Organometall. Chem. *23*, 547 (1970)
217) Ryan, R. C., Dahl, L. F.: J. Amer. Chem. Soc. *97*, 6904 (1975)
218) Saillant, R., Barcelo, G., Kaesz, H. D.: J. Amer. Chem. Soc. *92*, 5739 (1970)
219) Sartorelli, U., Canziani, F., Martinengo, S., Chini, P.: **XII**, I.C.C.C., (1970), p. 144
220) Shapley, J. R., Keister, J. B., Churchill, M. R., De Boer, B. G.: J. Amer. Chem. Soc. *97*, 4145 (1975)
221) Shore, N. E., Ilenda, C. S., Bergman, R. G.: J. Amer. Chem. Soc. *98*, 256 (1976)
222) Simon, G. L., Dahl, L. F.: J. Amer. Chem. Soc. *95*, 2175 (1973)
223) Simon, G. L., Dahl, L. F.: J. Amer. Chem. Soc. *95*, 2164 (1973)
224) Sisak, A., Sisak, C., Ungváry, F., Palyi, G., Markó, L.: J. Organometall. Chem. *90*, 77 (1975)
225) Skinner, H. A.: Adv. Organometall. Chem. *2*, 110 (1960) and references therein
226) Stalinski, B., Coogan, C. K., Gutowsky, H. S.: J. Chem. Phys. *34*, 1191 (1961) and references therein
227) Steele, W. V.: Bond energies, data and methodology. London: Butterworths 1974
228) Stuntz, G. F., Shapley, J. R.: Inorg. Nucl. Chem. Lett. *12*, 49 (1976)
229) Tables of interatomic distances, Spec. Publ., No. 18, The Chem. Soc., London, 1965
230) Terzis, A., Spiro, T. G.: Chem. Comm. **1970**, 1160
231) Tilney-Bassett, J. F.: J. Chem. Soc. **1963**, 4784
232) Todd, L. J., Wilkinson, J. R.: J. Organometall. Chem. *77*, 1 (1974)
233) Tolman, C.: J. Amer. Chem. Soc. *92*, 2956 (1970)
234) Trinh-Toan, Fehlhammer, W. P., Dahl, L. F.: J. Amer. Chem. Soc. *94*, 3389 (1972)
235) Ungváry, F., Markó, L.: Inorganica Chimica Acta *4*, 324 (1970)
236) Ungváry, F., Markó, L.: J. Organometall. Chem. *71*, 283 (1974)
237) Vollhardt, P. C., Bercaw, J. E., Bergman, R. G.: J. Amer. Chem. Soc. *96*, 4998 (1974)
238) Vollhardt, P. C., Bercaw, J. E., Bergman, R. G.: J. Organometall. Chem. *97*, 283 (1975)
239) Vranka, R. G., Dahl, L. F., Chini, P., Chatt, J.: J. Amer. Chem. Soc. *91*, 1574 (1969)
240) Wei, C. H.: Thesis, University of Wisconsin
241) Wei, C. H., Dahl, L. F.: J. Amer. Chem. Soc. *88*, 1821 (1966)

242) Wei, C. H., Wilkes, G. R., Dahl, L. F.: J. Amer. Chem. Soc. *89*, 4792 (1967)
243) Wei, C. H.: Inorg. Chem. *8*, 2384 (1969)
244) Wei, C. H., Markó, L., Bor, G., Dahl, L. F.: J. Amer. Chem. Soc. *95*, 4840 (1973)
245) Wender, I., Sternberg, H. W., Orchin, M.: J. Amer. Chem. Soc. *74*, 1216 (1952)
246) White, J. W., Wright, C. J.: J. Chem. Soc. (A) **1971**, 2843
247) Whyman, R.: Chem. Comm. **1969**, 1381
248) Whyman, R.: J. Organometall. Chem. *24*, C35 (1970)
249) Whyman, R.: J. Chem. Soc. Dalton **1972**, 2294
250) Whyman, R.: J. Chem. Soc. Dalton **1972**, 1375
251) Wilson, C. J., Green, M., Mawby, R. J.: J. Chem. Soc. Dalton **1974**, 421
251a) Wilson, R. D., Bau, R.: J. Amer. Chem. Soc., in press.
252) Yawney, D. B., Stone, F. G. A.: J. Chem. Soc. (A) **1969**, 502
253) Yawney, D. B. W., Doedens, R. J.: Inorg. Chem. *11*, 838 (1972)

Received July 5, 1976

Thermochemical Studies of Organo-Transition Metal Carbonyls and Related Compounds

Dr. J. A. Connor

Department of Chemistry, The University, Manchester M13 9PL, England

Table of Contents

Note: The bond enthalpy contributions (b.e.cs) are shown in the tables to the nearest whole kilojoule (kJ).

1. Introduction

The complaint is often made, in print and elsewhere, that the amount of thermochemical information concerning organometallic compounds of the transition metals is so meagre as to preclude sensible discussion of bond enthalpies in these compounds, and almost implying that these enthalpies can safely be ignored in favour of qualitative notions of the strength of bonds. For example, until recently there was a common, if unspoken, consensus of opinion that metal-carbon (alkyl) bonds were thermochemically "weak" in the absence of special "stabilising" factors, and the same "weak" description has been given to metal-hydrogen bonds. In the past the lack of information has occasionally been supplemented by estimates which were made on rather uncertain grounds. The result of this has been that the state of thermochemical knowledge is thought to be rather worse than is the case in fact. During the past five or six years an effort has been made by various groups to determine bond enthalpies of transition metal organometallic compounds and the true pattern is now becoming clearer.

It is not intended to discuss the details of the various methods of thermochemical measurement and the evaluation of results. This has been done in authoritative articles by Skinner[1] and by Pilcher[2] which have appeared recently, and which deal specifically with the thermochemistry of organometallic compounds. Instead this article will survey the results which may be derived from the information which is available and relate them to features of metal carbonyl chemistry in particular.

At the outset it is worth recalling some of the limitations which impede the rapid and comprehensive development of our knowledge of bond enthalpy contributions (b.e.cs) of the various element-ligand bonds formed by transition metals. First, very many of the organometallic compounds of these elements are available only in very small quantities. This problem of sample size may arise because of the cost of the basic raw material from which the desired compound might be made, or because of difficulties in the synthesis which result in low yields and wastage in purification, or finally because of the not inconsiderable problem of decomposition on storage. This last problem is often overlooked and may indeed be less important for synthetic or other purposes, but the fact that, for example, $[Mo(CO)_3(pyr)_3]$ slowly decomposes in a vacuum and on storage in a nitrogen atmosphere below ambient temperature, releasing pyridine and forming $Mo(CO)_6$, hinders the measurement of the true enthalpy of formation of the pure compound.

Secondly, the well-established calorimetric techniques (*e.g.* rotating bomb, hot zone) which are capable of achieving high precision are generally dependent upon the use of quantities of material greater than 0.1 g and preferably about 1 g for each measurement. A satisfactory determination of the heat of combustion of a metal-containing compound may very well require several measurements, ten measurements would not be regarded as exceptional. From this, the principle conflict between the experimental requirements and the availability of material is obvious. The development of a rotating micro-bomb calorimeter would be particularly beneficial to the study of compounds of the 4*d*- and 5*d*- transition metals.

While awaiting the development of precise combustion calorimetric techniques on a micro scale, much of the thermochemical information obtained during the past few years on transition metal organometallic compounds has come from the use of the high temperature microcalorimeter as a hot-zone device[3], to measure heats of thermal decomposition and of reaction with, for example, iodine vapour. This procedure has the advantages that only small samples (ca. 5 mg) are used in each measurement and that the two independent determinations (enthalpy of decomposition and enthalpy of reaction) can be compared directly. For example, less than 0.1 g of $Co_4(CO)_{12}$ was sufficient for measurements of the heat of decomposition and iodination[4], giving closely consistent values of ΔH_{298} by both methods which led to ΔH_f° [$Co_4(CO)_{12}$, c] = −1845 ± 16 kJ mol^{-1}. A similar small quantity of [$Cr(\eta$-$C_6H_5Cl)(CO)_3$] gave mean values of the enthalpy of formation of the compound of −448 kJ mol^{-1} (thermal decomposition) and −466 kJ mol^{-1} (iodination), of which the latter was preferred[5].

These examples draw attention to two important aspects of the microcalorimetric technique, namely accuracy and product identification.

1.1. Accuracy

Where it has been possible to compare directly the results of high precision hot zone or bomb calorimetric methods with those obtained by microcalorimetric measurements of the heat of decomposition or iodination, the agreement has generally been good. For example, ΔH_f° [$Cr(CO)_6$, c] = −980 (conventional bomb)[6], −982 (microcalorimeter; iodination[3]) kJ mol^{-1}, or ΔH_f° [$Cr(\eta$-$C_6H_6)_2$, c] = + 146 (combustion[7]), + 142 (microcalorimeter; iodination[8]) kJ mol^{-1}. These results suggest that one may be cautiously optimistic about the accuracy of measurements made on more complex compounds, such as those containing metal clusters or ligands which are likely to be reactive when released from the metal at high temperatures. However, while values determined with high precision are always to be desired and preferred, it is important to understand that much useful information can be obtained from measurements of lower precision. For example, very little is to be gained in terms of improving the precision of the derived b.e.cs by a more exact measurement of the enthalpy of formation of $Rh_6(CO)_{16}$ for which the microcalorimetric (thermal decomposition) value[9] ΔH_f° [$Rh_6(CO)_{16}$, c] = −2418 ± 17 kJ mol^{-1}. A change of 150 kJ mol^{-1} (6.2 per cent) in ΔH_f° would change each b.e.c by 5 kJ mol^{-1}, which is hardly significant either in terms of the error or the general application of such b.e.c values.

1.2. Product Identification

Combustion of transition metal organometallic compounds produces a mixtures of simple compounds (metal oxides, carbon oxides, water, nitrogen) which is subject to exact analysis. Thermal decomposition or high temperature iodination of the same compounds cannot necessarily be expected to produce simple materials, with the result that identification is often a difficult problem. This is typified by diene derivatives of iron carbonyl[10], where side reactions of the dienes (*e.g.* polymerization) follow disruption of the iron-diene bonds. The oligomeric mixture can be partially analysed by mass spectrometry. By assuming that dimerization occurs in the primary disruption process, it is then possible to derive a value for the enthalpy of disruption.

In general, although the results of microcalorimetric studies do not pretend to provide enthalpy values of the very highest accuracy presently attainable by macroscale combustion calorimetry, they do offer a basis for application on a wide scale and are sufficiently precise for most purposes. The conflicting claims of accuracy and usefulness are particularly acute in the area of transition metal organometallic chemistry: this review will attempt to follow a middle way between them.

1.3. Transferability

This assumption is crucial to the development of a body of thermochemical information on compounds of the transition metals. The assumption is made that b.e.c values are transferable from one compound to another in which the formal oxidation state of the metal is unchanged. This means that, for example, the value of $\bar{D}$(Cr-CO) in $Cr(CO)_6$ is assumed unchanged[a)] in $[Cr(\eta\text{-}C_6H_6)(CO)_3]$, or that $\bar{D}(Cr\text{-}C_6H_6)$ in the same compound is assumed unchanged from its value in $Cr(\eta\text{-}C_6H_6)_2$. In this example, $\Delta H_f^\circ\,[Cr(\eta\text{-}C_6H_6)(CO)_3, q] = -352.3$ kJ mol^{-1}, from which the enthalpy of disruption, ΔH_D, for the process

$$[Cr(\eta\text{-}C_6H_6)(CO)_3]\,(g) \rightarrow C_6H_6\,(g) + Cr(g) + 3\,CO\,(g)$$

can be calculated to be 502 kJ mol^{-1}. The value[6)] of $\bar{D}$(Cr-CO) in $Cr(CO)_6$ (see Section 2.1.) is 108 kJ mol^{-1}, so that $\bar{E}(Cr\text{-}C_6H_6) = 178$ kJ mol^{-1}. Alternatively, the value of $\bar{D}(Cr\text{-}C_6H_6)$ found[7, 8)] in $Cr(\eta\text{-}C_6H_6)_2$ (see Section 2.3.) is 163 kJ mol^{-1} so that $\bar{E}$(Cr-CO) = 113 kJ mol^{-1}. In view of the experimental error in the measurements from which the b.e.c values are derived (±12 kJ mol^{-1}) the assumption of transferability seems justified, but the limits within which it is valid have yet to be determined. An indication of possible limitations on the assumption is provided by a comparison of $[Cr(\eta\text{-}B_3N_3Me_6)(CO)_3]$, which decomposes thermally at temperatures below 420K, with $[Cr(\eta\text{-}C_6Me_6)(CO)_3]$ which undergoes thermal decomposition only at temperatures above 600K. It would be reasonable to propose that the (Cr-CO) b.e.c in these two compounds cannot be the same and that the undoubted enhancement in the binding enthalpy of $[Cr(\eta\text{-}C_6Me_6)(CO)_3]$ must be located to some extent in the Cr-CO bonds, but without information about the value of $\bar{D}(Cr\text{-}B_3N_3Me_6)$ in the hypothetical $[Cr(\eta\text{-}B_3N_3Me_6)_2]$ it is not possible to refine or improve the assumption of the transferability of the Cr-CO b.e.c at this time.

---

a) Here we are introducing two symbols $\bar{D}, \bar{E}$ without prior definition. In $Cr(CO)_6$, $\bar{D}$(Cr-CO) = $\Delta H_D/6$. By transferring this $\bar{D}$ to $[Cr(\eta\text{-}C_6H_6)(CO)_3]$ unchanged, the resultant $\bar{E}(Cr\text{-}C_6H_6)$ depends on $\bar{D}$(Cr-CO). It is distinguished by an $\bar{E}$ value because of its dependency on an assumed transferable $\bar{D}$.

2. Binary Compounds, ML_n

2.1. Metal Carbonyls, L = CO

Many of the mononuclear metal carbonyls have been the object of detailed and precise calorimetric measurement by Wilkinson[10] and Skinner[6] and their respective groups. Measurements on the more complex, polynuclear carbonyls have been made almost exclusively by microcalorimetry[4, 9], with the notable exception of Good's work[12] on $Mn_2(CO)_{10}$. The results of these measurements are collected in Table 1.

Table 1. Standard enthalpy of formation of metal carbonyls $[M_m(CO)_n]$ in the gas phase. Bond description and bond enthalpy contributions ($\bar{T}$, $\bar{M}$ and $\bar{B}$) to the enthalpy of disruption, ΔH_D

All values given in kJ mol^{-1}

Compound	ΔH_f° (g)	Ref.	ΔH_D	Description	$\bar{T}$	$\bar{M}$	$\bar{B}$
$Cr(CO)_6$	−908.3 ± 2.0	6)	646	$6T$	108		
$Mo(CO)_6$	−916.2 ± 1.8	13)	910	$6T$	152		
$W(CO)_6$	−884.5 ± 2.5	14)	1069	$6T$	178		
$Mn(CO)_5$	−767.8 ± 6	15)	496	$5T$	99		
$Mn_2(CO)_{10}$	−1597.5 ± 5.3	3, 12, 16)	1068	$10T+M$	100	67	
$Re(CO)_5$	686.2 ± 6	17)	908	$5T$	182		
$Re_2(CO)_{10}$	−1559.4 ± 21	4, 16, 18)	2029	$10T+M$	187	128	
$Fe(CO)_5$	−723.8 ± 6.4	11)	585	$5T$	117		
$Fe_2(CO)_9$	−1335 ± 25	4)	1173	$6T+6B+M$	117	82	64
$Fe_3(CO)_{12}$	−1753 ± 28	4)	1676	$10T+4B+3M$	117	82	64
$Ru_3(CO)_{12}$	−1820 ± 28	4)	2414	$12T+3M$	172	117	
$Os_3(CO)_{12}$	−1644 ± 28	4)	2666	$12T+3M$	190	130	
$Co(CO)_4$	−561 ± 12	2)	544	$4T$	136		
$Co_2(CO)_8$	−1172 ± 10	4, 19)	1160	$6T+4B+M$	136	83	68
$Co_4(CO)_{12}$	−1749 ± 28	4)	2130	$9T+6B+6M$	136	83	68
$Rh_4(CO)_{12}$	−1749 ± 28	4)	2649	$9T+6B+6M$	166	114	83
$Rh_6(CO)_{16}$	−2299 ± 28	9)	3496	$12T+8B+11M$	166	114	83
$Ir_4(CO)_{12}$	−1715 ± 26	4)	3051	$12T+6M$	190	130	
$Ni(CO)_4$	−600.4 ± 4.0	20)	588	$4T$	147		

The values of ΔH_f° (g) carry both the experimental uncertainty in the standard enthalpy of formation of the crystalline (or liquid) metal compound and the uncertainty (experimental or estimated) in the enthalpy of sublimation (or vaporization).

For a mononuclear carbonyl, $M(CO)_n$, the enthalpy of disruption ΔH_D refers to the process

$$M(CO)_n[g, 298K] \rightarrow M[g, 298K] + nCO[g, 298K]$$

and is calculated from

$$\Delta H_D = \Delta H_f^\circ[M, g] + n\Delta H_f^\circ[CO, g] - \Delta H_f^\circ[M(CO)_n, g]$$

where [M, g] implies metal atoms in their ground states. The metal-ligand bonds in $M(CO)_n$ are all of the *terminal* type. The mean enthalpy of dissociation of these bonds is symbolized by $\overline{T} = \Delta H_D/n$ in Table 1. In a polynuclear carbonyl, $M_m(CO)_n$, the quantity ΔH_D refers to the process

$$M_m(CO)_n[g, 298K] \rightarrow mM[g, 298K] + nCO[g, 298K]$$

In addition to terminal metal-carbonyl ligand bonds, a polynuclear metal carbonyl contains metal-metal bonds and, perhaps, bridging metal-carbonyl linkages as well, both of which will make an enthalpy contribution to ΔH_D which is symbolized by $\overline{M}$ (for *metal*-metal) and $\overline{B}$ (for *bridging* metal-carbonyl) in Table 1. The structures of most mono- and poly-nuclear binary carbonyls are known[21] and the bond descriptions in Table 1 assume that the crystal molecular structures are retained in the gaseous phase. Insofar as mass spectrometry may show that molecular ions are formed from these compounds, it seems reasonable to suppose that the structure does not change significantly. It is also pertinent to recall recent work[22] on the dynamic structure of polynuclear carbonyls in solution, which shows that there is rapid scrambling of CO ligands.

The values of $\overline{B}$ and $\overline{M}$ in the carbonyls of iron and cobalt were calculated by assuming that the value of $\overline{T}$ in $Fe(CO)_5$ and in $[\cdot Co(CO)_4]$ is transferable and then solving the simultaneous equations as in the case of iron,

$Fe(CO)_5$	$5\overline{T}$	$= 585$
$Fe_2(CO)_9$	$\overline{M} + 6\overline{B} + 6\overline{T}$	$= 1173$
$Fe_3(CO)_{12}$	$3\overline{M} + 4\overline{B} + 10\overline{T}$	$= 1676$

From the values of $\overline{T}$, $\overline{M}$ and $\overline{B}$ calculated in this way we may note that for both iron and cobalt $\overline{M} \sim 0.68\,\overline{T}$ and that $\overline{B} \sim 0.5\,\overline{T}$. The latter approximation is intuitively consistent with the observations[22] of bridge-terminal exchange of coordinated CO. If we now assume that these empirical approximations apply to other polynuclear carbonyls, it is possible to extract the $\overline{T}$ and $\overline{M}$-values for metals other than iron and cobalt, despite the limited ΔH_D data which are available.

If the enthalpy of sublimation (atomization) of the metal, $\Delta H_f(M, g)$ is considered, these empirical relationships can be extended, if it is assumed that the major contribution to the cohesive energy of the metal originates in the metallic bonding between *adjacent* metal atoms in the lattice so that each M-M bond contributes $2\Delta H_f^\circ(M, g)/n$ kJ mol^{-1}, where n is the coordination number of the metal M in the bulk metal. The dissociation energies of various transition metal diatomic molecules, $D_0(M_2)$/kJ mol^{-1} are shown in Table 2. The metallic bond is electron deficient with respect to a normal covalent diatomic M-M bond in M_2, which might therefore be expected to be both shorter and stronger. Evidence in support of this is provided by the observation that the dissociation energy of $Rh_2(g)$ is reported[24] to be

Table 2. Dissociation energies of diatomic transition metal. molecules, D_0(M-M) kJ mol^{-1}

M	D_0(M-M)	M	D_0(M-M)
Sc	159	Y	155
Ti	125	Rh	276
V	238	Pd	63
Cr	151	Ag	161
Mn	42	Au	221
Fe	100		
Co	167		
Ni	258		
Cu	197		

Values taken from Kondratiev [23], except for Ni, Rh and Au (Ref. [24]).

276 ± 25 kJ mol^{-1} with an estimated bond length of 228 pm, which is much shorter than the interatomic separation in the metal, 269 pm. Consequently we may write

$$\bar{M} \sim 2\,[\Delta H_f^\circ(\text{M, g})]/n \qquad (1)$$
$$\bar{T} \sim (0.28 \pm 0.04)\,[\Delta H_f^\circ(\text{M, g})] \qquad (2)$$

The basis of Eq. (2) is shown in Fig. 1, where values of $\bar{T}$ are plotted against ΔH_f° (M, g).

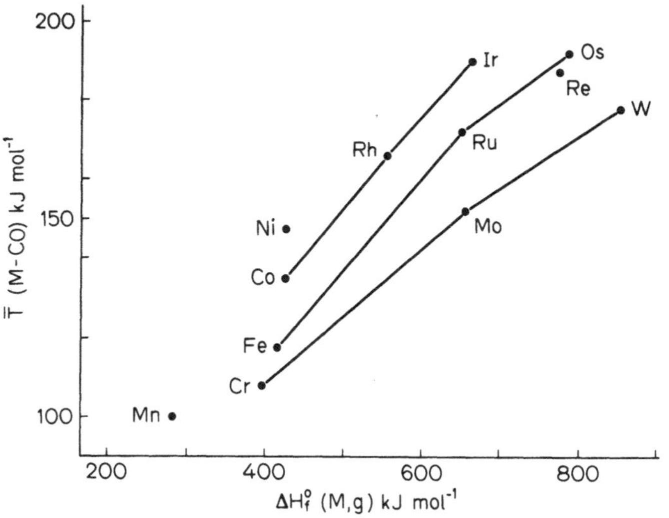

Fig. 1. Variation of the terminal metal-CO bond enthalpy contribution, $\bar{T}$ (M-CO) kJ mol^{-1} as a function of the enthalpy of atomization of the metal, ΔH_f° (M, g) kJ mol^{-1}

Most of the metals which form *polynuclear* carbonyls adopt the 12 co-ordinate face-centred cubic structure ($n = 12$) in the solid state so that the relation between $\bar{M}$ and $\bar{T}$ can be revised to

$$\bar{M} \sim 0.6\bar{T} \qquad (3a)$$

and, for 8 co-ordinate metals

$$\bar{M} \sim 0.9\bar{T} \qquad (3b)$$

Table 3. Standard enthalpy of formation of the gaseous metal atom, ΔH_f° (M, g), coordination number n of the atom in the bulk metal at room temperature and values of $\bar{M}$ and $\bar{T}$ calculated by equations 1 and 3. Thermochemical values in kJ mol^{-1}

	Metal	ΔH_f° (M, g)[1]	n[2]	$\bar{M}$ (calc)[3]	$\bar{T}$ (calc)[4]
3d-Series	Sc	377.8	12c	63	105
	Ti	469.9	12h	78	131
	V	514.6	8	129	143
	Cr	397.5	8	99	110
	Mn	284.5	8 [5]	71	79
	Fe	417.1	8	104	116
	Co	428.4	12h	71	119
	Ni	429.3	12c	71	119
	Cu	337.6	12c	56	94
4d-Series	Y	424.7	12h	71	118
	Zr	608.8	12h	102	169
	Nb	722.6	8	181	201
	Mo	656.9	8	164	182
	Tc	695.0	12h	116	193
	Ru	651.0	12h	109	180
	Rh	557.3	12c	93	155
	Pd	372.4	12c	62	103
	Ag	284.9	12c	47	79
5d-Series	La	430.9	12h	72	120
	Hf	619.2	12h	103	172
	Ta	786.6	8	197	219
	W	853.5	8	213	237
	Re	775.7	12h	129	215
	Os	789.9	12h	132	219
	Ir	665.2	12c	111	185
	Pt	567.7	12c	94	157
	Au	368.8	12c	61	102

1) Values taken from Ref.[2] and[23].
2) Data taken from Ref.[25]; h = hexagonal, c = cubic.
3) $\bar{M} = 2\,[\Delta H_f^\circ(\text{M, g})]/n$.
4) $\bar{T} = 1.667\,\bar{M}$ for $n = 12$; $\bar{T} = 1.111\,\bar{M}$ for $n = 8$.
5) The structure of α-manganese (Ref.[26]) also shows 12-coordinate Mn atoms for which $\bar{M} = 47$ and $\bar{T} = 80$.

Table 3 shows the values of $\bar{M}$ and $\bar{T}$ evaluated for all the transition metals in terms of Eq. (1) and (3). Comparison of these estimates with the values in Table 1 shows satisfactory agreement in most *polynuclear* systems. While it is *not* suggested that there is any necessary relation between $\bar{M}$, $\bar{T}$ and ΔH_f° (M, g), it is clear that these empirical relationships provide a useful index of the strengths of bonds in polynuclear systems and give at least some indication of the magnitude of b.e.cs in other systems.

For example, the results in Table 3 suggest that binary carbonyls of copper, silver and gold which have been detected spectrometrically in matrices at very low temperatures[27], contain metal-CO bonds which are approximately of the same strength as those in $Mn_2(CO)_{10}$. Similar considerations apply to carbonyls of palladium and platinum which have also been detected by matrix isolation spectrometry[28]. All of these binary compounds are unstable with respect to [M(c) + CO(g)] at room temperature.

The principal conclusions to be drawn from the b.e.cs of metal carbonyls, which are of more importance than the values themselves, are these:

a) The metal-CO bond enthalpy ($\bar{T}$) is always greater than the metal-metal bond enthalpy ($\bar{M}$) for a given metal. This bears with it the implication that, at least for small clusters the energy of the metal-CO interaction dominates any effect resulting from the energy of the metal-metal interaction. This provides some support for the proposal[29] that the structures of polynuclear metal carbonyls should be seen in terms of fitting a particular cluster of metal atoms into the vacancies left in a regular geometric arrangement of carbon monoxide molecules. It is possible to speculate that for larger polymetallic clusters such as $[Rh_{12}(CO)_{30}]^{2-}$, the energy of the M-CO interaction will decrease in importance relative to that of the M-M interaction, but there is no real basis for this at present. It is worth pointing to an important limitation of this picture in terms of its application to enthalpies of CO adsorption on bulk metal surfaces. The enthalpy of adsorption (whether initial or integral) of CO on metals may include within it dissociative adsorption (as in the case of the earlier members of a transition series) as well as associative adsorption (as in the later members of a transition series). Measurements of the oxygen ls binding energy of carbon monoxide on metals show an inverse relationship with the enthalpy of adsorption[30]. This is consistent with dissociative adsorption (to give C + O) on metals which show large enthalpies of adsorption. For this reason, there is no necessary correlation between the adsorption enthalpy[31] (Table 4) and the value of $\bar{T}$ for the same metal. Recent studies using ultraviolet photoelectron spectroscopy have demonstrated[37] that there are close similarities between the CO of $Rh_6(CO)_{16}$ and carbon monoxide absorbed on a Pd(111) single crystal surface. The metal-CO bond formed on chemisorption therefore has a localized character and this provides an important justification for the use of cluster compounds of the later metals of a transition series as models of metal surfaces for chemisorption and catalysis.

b) For a given metal, both $\bar{T}$ and $\bar{M}$ are small compared with b.e.c. values (especially $\bar{D}$ values) for other ligands such as halides. (See also under Section 2.2. and Table 26).

c) There is no purely thermodynamic reason why neutral binary carbonyl compounds of metals such as zirconium, tantalum and platinum should be unstable with respect to decomposition to the metal and carbon monoxide at ambient tem-

Table 4. Initial (A) and integral (B) enthalpy of adsorption of CO on polycrystalline evaporated metal films (kJ mol^{-1}) (Ref.[31])

3*d*-Series	A	B	4*d*-Series	A	B	5*d*-Series	A	B
Ti	640	628	Zr	632	619	Ta	561	536
Mn	326	318	Nb	552	485	W	527	335
Fe	192	146	Mo	310	251	Pt	201	184
Co	197	192	Rh	192	184			
Ni	176	167	Pd	180	167			

Heat of adsorption (Q, kJ mol^{-1}) of CO on single crystals of f.c.c. metals

Metal	Face	Q	Ref.
Ni	100	126	32)
	110	126	32)
	111	111	32)
Cu	111	50	32)
Ru	001	121	33)
	101	118	114)
Pd	111	142	34)
Ag	111	27	32)
Ir	110	155	35)
	111	146	36)
Pt	111	146	32)

perature. The fact that compounds such as $[Cp_2Zr(CO)_2]$ ($Cp = \eta$-C_5H_5) (Ref.[38]) and $[CpTa(CO)_4]$ (Ref.[39]) have been synthesized, albeit with some difficulty, and that stable anionic species such as $Nb(CO)_6^-$ (Ref.[40]) and $[Pt_3(CO)_3(\mu_2\text{-}CO)_3]_n^{2-}$ (n = 2–5) (Ref.[41]) are known, is to be seen as supporting this argument.

The intrinsic enthalpy change in the reaction

$$M(CO)_n[g, 298K] \rightarrow M^* [298K] + nCO^* [298K]$$

in which M* and CO* represent the *valence* states of the metal and carbon monoxide respectively, is not represented by $n\bar{T}$, as this fails to take account of the valence state reorganization energy of M and the molecular electron reorganization energy of CO. Estimates of the factors contributing to the bond enthalpy of the M-CO link were first made by Cotton, Fischer and Wilkinson[11] for M = Cr, Fe and Ni. More recently, the best available data has been reassessed[42] for the 3*d*-transition metals Cr-Ni. The results which are shown in Table 5 relate to the Scheme. The effectively constant value (365 ± 10 kJ mol^{-1}) of the intrinsic M-C enthalpy ΔH^* for these five metals shows that the metal ligand bonding is qualitatively the same in all the 3*d*-transition metal carbonyls. This is in agreement, not only with the conclusions

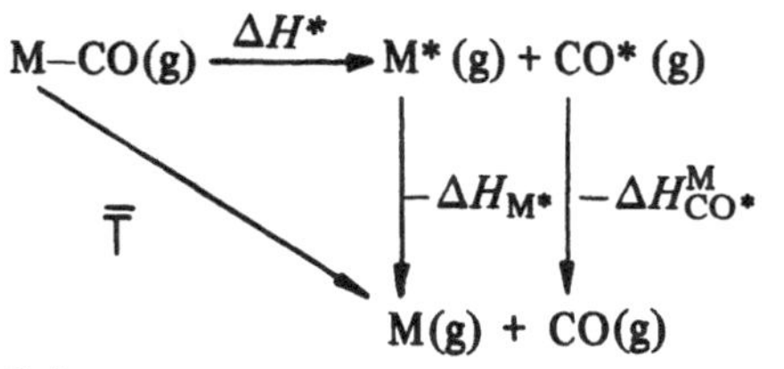

Scheme

Table 5. Contributions to the enthalpy of disruption to valence state, ΔH^* kJ mol^{-1} of the M-CO bond

M	$\bar{T}$	$\Delta H^{M}_{CO^*}$	ΔH_{M^*}	ΔH^*
Cr	108	94	159	361
Mn	99	95	167	361
Fe	118	80	177	375
Co	136	61	169	366
Ni	147	49	165	361

$\bar{T}$ – mean enthalpy of disruption to ground state products;
$\Delta H^{M}_{CO^*}$ – valence reorganization energy of CO;
ΔH_{M^*} – valence state promotion energy of M;
$\Delta H^* = \bar{T} + \Delta H^{M}_{CO^*} + \Delta H_{M^*}$.

of molecular orbital calculations[43)] but also with measurements of the Cls and Ols core electron binding energies which are effectively constant for these neutral binary carbonyls[44)]. The value of ΔH^* for $4d$ and $5d$ transition metal carbonyls is estimated[42)] as approximately 415 and 460 kJ mol^{-1} respectively.

Mention was made earlier that platinum does not form a neutral binary carbonyl which is stable at ambient temperature. In contrast to this, the known stability[45)] of $Pt(PF_3)_4$ together with mass spectroscopic determination[46)] of the mean metal-PF_3 bond dissociation enthalpy in $Ni(PF_3)_4$ which indicated that this was significantly larger than the corresponding Ni-CO bond enthalpy, has led to the implication that the M-PF_3 bond is stronger than the analogous M-CO bond. That this assumption is incorrect is shown by the observation[47)] that $\bar{D}$ (M-PF_3) in $Cr(PF_3)_6$ (107 kJ mol^{-1}) and $Ni(PF_3)_4$ (147 kJ mol^{-1}) are the same as the values of $\bar{T}$ (see Table 1). Thermodynamic arguments regarding the particular stability of certain trifluorophosphine complexes are therefore unjustified. Instead, a kinetic argument based on the greater ability of the trifluorophosphine ligand to shield the metal from nucleophilic attack by virtue of its greater size than the rod-like carbon monoxide ligand seems appropriate. This is made clear when models of $Cr(PF_3)_6$ and $Cr(CO)_6$ are considered; the former is a spherical molecule with the metal buried at the centre of a shell provided by the regularly disposed fluorine atoms. This example demonstrates some of the dangers which beset the use of mass spectroscopy to determine bond enthalpies in systems of this kind. In general, if the excitation energy and excess kinetic energy of the ions produced by electron impact are neglected then the measured appearance potential can only give an upper limit to the value of the bond dissociation enthalpy. Where binary metal carbonyls themselves are concerned, electron impact determinations of $\bar{T}$ give values which are in the correct sequence for M-CO (M = Cr, Mo, W, Fe, Ni), but even the lowest recorded[48] values of the appearance potentials of these molecules lead to values of $\bar{T}$ which are approximately 20 kJ mol^{-1} greater than those in Table 1.

2.2 Metal Alkyls, L = Alkyl

Progress with measuring bond enthalpies of this class of compound[49] has come only recently and information is restricted to the metals of groups 4 to 6. In general, these measurements have been made by reaction calorimetry in solution; the homoleptic metal alkyl is hydrolyzed to give the metal oxide[50] or reacted with an alcohol to give the metal alkoxide[51]. Solution calorimetry in these systems is complicated, because the products of acid hydrolysis are pH-dependent. The choice of isopropanol[51] for the alcoholysis of the group 4 alkyls was made to ensure that the product was a monomeric alkoxide. It is instructive to compare the mean bond enthalpy obtained for homoleptic alkyls with other homoleptic compounds of the same metals

Table 6. Mean bond dissociation enthalpy $\bar{D}$ (M-L) kJ mol^{-1} for homoleptic metal compounds ML_n

	M = Ti	Zr	Hf	Nb	Ta	Mo	W
L	n = 4	4	4	5	5	6	6
CH_3	(260)	(310)	(330)		261		159
CH_2CMe_3	188	227	224				
CH_2Ph	203	252					
NMe_2	307	350	(370)		328		220
NEt_2	309	343	367				
OMe				419	439		(360)
OEt	429			417	440		
OPr^n	441			418	440		
OPr^i	444	518	534				
OBu^t	434						
F	586	648	665	573	603	449	508
Cl	430	490	497	406	430	305	347

Data for MF_n and MCl_n from Ref.[23]; for ML_4[M=Ti, Zr, Hf, L=CH_2CMe_3, CH_2Ph, NMe_2, NEt_2, OPr^i] from Ref.[51]; for $Ti(OR)_4$ (R=Et, Pr^n, Bu^t) from Ref.[52]; for $M(OR)_5$ (M=Nb, Ta, R=Me, Et, Pr^n) from Ref.[53]; for $M(NMe_2)_n$ (M=Ta, n = 5; M=W, n = 6) from Ref.[54]; for $M(CH_3)_n$ (M=Ta, n = 5; M=W, n = 6) from Ref.[50]. The values of $\bar{D}$ (M-L) are from, or recalculated from, the references cited, using ΔH_f° (M, g) values in Table 3 and the enthalpies of formation ΔH_f° (L, g) taken from Refs.[2] and[23]. The values of $\bar{D}$ (M-L) given in brackets are estimates based on comparison with main group elements, (Ref.[2]) or extrapolation from other values in the table.

in which the ligands are halogenide, alkoxide and dialkylamide groups. These mean b.e.cs are shown in Table 6 as $\bar{D}$ values, that is, relating to the process

$$ML_n\,(g) \rightarrow M\,(g) + n\,L\,(g)$$

$$\text{where } \bar{D} = \Sigma[\Delta H_f^\circ\,(M, g) + n\Delta H_f^\circ\,(L, g) - \Delta H_f^\circ\,(ML_n, g)]/n$$

An alternative method of calculation produces $\bar{E}$-values. The important difference between these two methods of calculation is that $\bar{D}$ values automatically incorporate the difference between benzyl and neopentyl substituents which is not the case for $\bar{E}$ values. That is to say that the $\bar{D}$ method starts out from the experimentally measured bond dissociation energies of the alkanes [D($PhCH_2$-H) = 365.1 kJ mol^{-1}; $\bar{D}$(Me_3CH_2-H) = 416.7 kJ mol^{-1}; D(CH_3-H) = 435.1 kJ mol^{-1}]. In the $\bar{E}$ method one starts out from an assumed constant value for E(C-H) and introduces corrections to take account of steric and electronic effects. Both methods can give the same result for a dissociation process, but $\bar{E}$ (C-M) for a benzyl-metal bond would not be the same as D(benzyl-M) unless allowance is made for the reorganization energy contribution of the benzyl radical, which is large compared to that of the methyl radical. The diagrams (Figs. 2a and 2b) show the variation of $\bar{D}$ (M-L) with the type of ligating atom, L. In general, the mean bond enthalpy $\bar{D}$ (M-L) decreases in the order L = F > OR > Cl > NR_2 > CH_2R, and the values increase monotonically

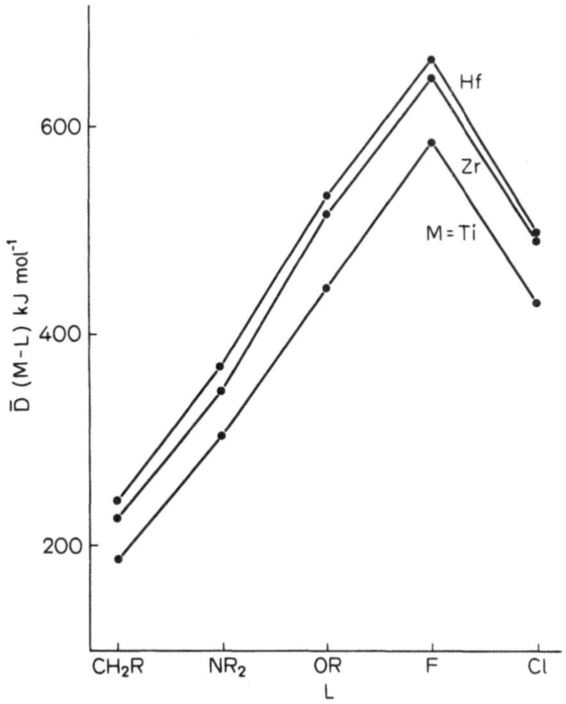

Fig. 2A. Variation of the mean bond dissociation energy, $\bar{D}$ (M-L) kJ mol^{-1}, in ML_4 (M = Ti, Zr, Hf) as a function of the group, L (L = CH_2CMe_3, NMe_2, OPr^i, F, Cl). see Table 6 and Ref.[51])

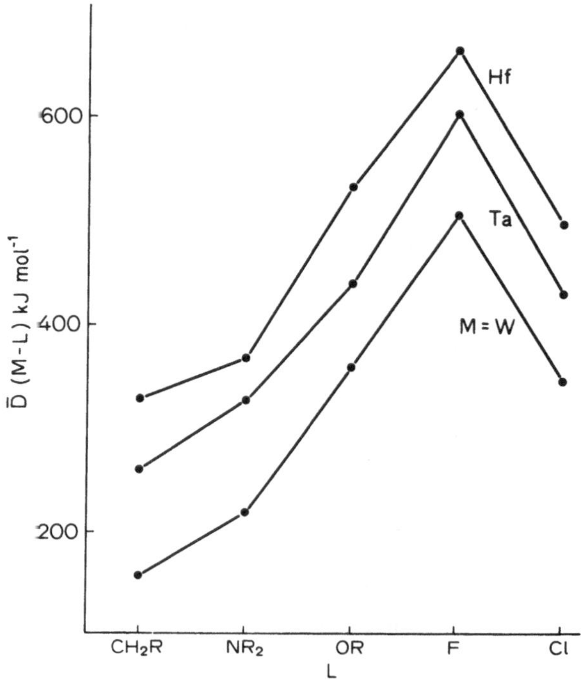

Fig. 2B. Variation of the mean bond dissociation energy, $\bar{D}$(M-L) kJ mol^{-1}, in ML_n (M = Hf, n = 4; M = Ta, n = 5; M = W, n = 6) as a function of the group L (L = CH_3, NMe_2, OMe, F, Cl) see Table 6 and Refs.[50, 51, 53, 54]. Note: The points corresponding to (Hf-CH_3), (Hf-NMe_2), and (W-OMe) are estimates

from one transition series to another in the order M(3d) < M (4d) < M (5d). The values of $\bar{D}$(M-OR) is remarkably insensitive to the nature of the substituent group, R. It has been suggested[52] that electronic factors, which might serve to strengthen the M-O bond through an increase in the ($p \rightarrow d$) π- bonding (Bu^t > Me), are counterbalanced by steric effects which will act to weaken the M-O bond.

It would appear that the mean bond enthalpy, $\bar{D}$(M-CH_2R) increases with increasing atomic number in any one group whereas in the main group (s, p-block) metals the mean bond enthalpy decreases in the same sense[2]. This observation is placed in a different perspective when $\bar{D}$(M-CH_2R) is compared with the standard enthalpy of formation of the gaseous atom, ΔH_f°(M, g). This shows (Fig. 3) that for both A- and B- sub-group metals, $\bar{D}$(M-CH_2R) increases with increasing ΔH_f°(M, g): this emphasizes the linear relationship between M-C and M-M bond enthalpies.

These data for $\bar{D}$(M-L) offer some basis for making predictions about the enthalpies of metal-carbon bonds involving other metals in these groups. It is important to bear in mind that all of the data in Table 6 concern the metals in their highest formal oxidation state. It is usually true that the mean bond enthalpy increases as the formal oxidation state of the metal decreases. This is exemplified by the values of $\bar{D}$(M-Cl) (M = Nb, Ta, Mo, W) in various oxidation states of M (Table 7a), and

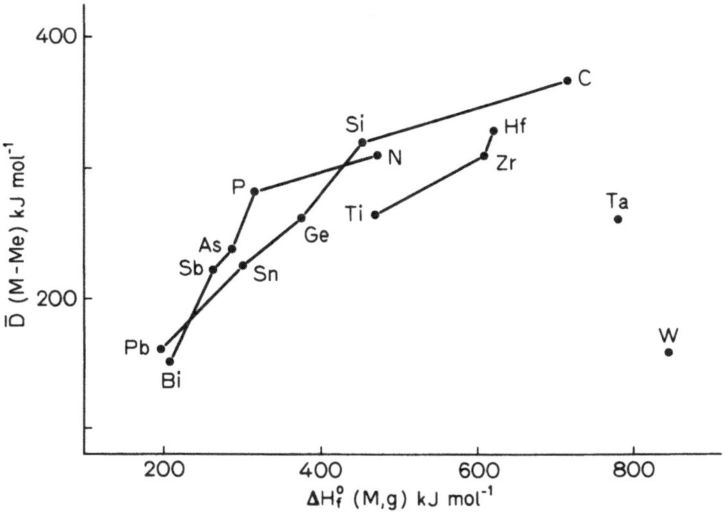

Fig. 3. Variation of the mean bond dissociation energy, $\bar{D}$ (M-Me) kJ mol^{-1} in MMe_n (m = C, Si, Ge, Sn, Pb; Ti, Zr, Hf, n = 4; M = N, P, As, Sb, Bi, n = 3; M = Ta, n = 5; M = W, n = 6) as a function of the enthalpy of atomization of the element, M, ΔH_f° (M, g) kJ mol^{-1}. Data for C-Pb and N-Bi from Ref.[2]; for Ti, Zr, Hf, Ta, W see Tables 3 and 6

by the variation of $\bar{D}$(M-OBu^t) (M = Ti, V, Cr) for the same formal oxidation state and coordination number of M (Table 7b). One might, therefore, expect that $\bar{D}$(M-CH_2R) will also increase as the formal oxidation state of M decreases. This may reflect the increase in electron density on the metal, and it is noteworthy that alkyl compounds of the heavier metals (4*d* and 5*d*) of group eight (*e.g.* Me_3PtI,

Table 7a. Variation of mean bond enthalpy, $\bar{D}$ (M-Cl) kJ mol^{-1} with formal oxidation state, (*d*-electron configuration) of the metal. (Ref.[23])

	d^n	M = Nb	Ta	Mo	W
n =	0	406	431	305	347
	1	439	456	360	376
	2	477		385	423

Table 7b. Mean bond enthalpy, $\bar{D}$ (M-OBu^t) kJ mol^{-1} in M(OBu^t)$_4$ (Ref.[55])

M	d^n	$\bar{D}$
Ti	0	434
V	1	366
Cr	2	305

$Pt(PPh_3)_2Me_2)$[b] have significantly greater thermal stability that $TiMe_4$, for example. At this point it is worth recalling that $\bar{D}$(W-CO) (Table 1) is only slightly greater than $\bar{D}$(W-CH_3) (178 compared with 159 kJ mol^{-1}), but the natures of the metal-ligand interaction in these two bonds differ greatly from one another.

In general, if $\bar{D}$(M-F) in MF_n is low then it is unlikely that the compound MMe_n could be thermodynamically stable. In group five, $\bar{D}$(V-F) = 468 kJ mol^{-1} (in VF_5)[23] which is much less than $\bar{D}$(Ta-F) in TaF_5 (Table 6), so that VMe_5 is expected to be insufficiently stable to exist at ambient temperatures. This is corroborated by the report[56] that $NbMe_5$ is less stable thermally than the tantalum analogue. The formal oxidation state of the metal is of great importance, as indicated by the occurrence of several homoleptic alkyls of vanadium(IV); the formation of anionic species also leads to an increase in stability[49]. The paramagnetic compound $ReMe_6$ has been described[57]; the value of $\bar{D}$(Re-F) in ReF_6 [436 kJ mol^{-1} (Ref.[58])] suggests that $\bar{D}$(Re-CH_3) may be in the range 75–90 kJ mol^{-1}, which is substantially less than the (Re-CH_3) b.e.c[16] in $CH_3Re(CO)_5$, a compound of rhenium(I) (see Section 3.2. and Table 13). In group eight, the validity of the extrapolation is much less certain in terms of the ratio $\bar{D}$(M-F)/$\bar{D}$(M-C), but the same principles are likely to apply. The low value of $\bar{D}$(Ru-F) in RuF_5 (< 370 kJ mol^{-1}, based on $\Delta H_f^\circ(RuF_5, c) = -892.9$ kJ mol^{-1} (Ref.[59])) would suggest that alkyl compounds of ruthenium(V), and $RuMe_5$ in particular, are unlikely to be thermally stable. Lastly, the fact that $\bar{D}$(U-F) in UF_6 [525 kJ mol^{-1} (Ref.[23])] is greater than $\bar{D}$(W-F) in WF_6 (Table 6) leads to the expectation that UMe_6 will be a thermally stable compound. The preparation of thermally stable, homoleptic anionic alkyls of uranium(IV), such as $[Li_2U(CH_2SiMe_3)_6 \cdot (tmen)_7]$ has been reported[60].

Mean bond enthalpy values cannot, of course, be directly correlated with the experimentally observed thermal stability of peralkyls. In the case of the group six metal carbonyls, $M(CO)_6$, although both $\bar{T}$ and the metal-carbon bond stretching force constant increase from M = Cr to M = W, the thermodynamic stability of $W(CO)_6$ with respect to equilibrium thermal decomposition is the least owing to the high enthalpy of sublimation of tungsten metal. On this basis, $Mo(CO)_6$ has the highest *thermodynamic* stability[61]. Where the metal alkyls are concerned, it is probable that the energy required to initiate decomposition is rather less than $\bar{D}$(M-CH_2R) and hydride transfer may be important even for the methyl group. That this is possible is demonstrated by the formation of $[Ta(CH_2CMe_3)_3(CHCMe_3)]$ (Ref.[62]) and $[Cp_2Ta(CH_2)Me]$ Ref.[63]. A quantitative indication of the extent to which initial decomposition may be favoured can be obtained from the stepwise dissociation enthalpies of some metal chlorides (Table 8). These show that the difference, Δ, between D_1 and $\bar{D}$ increases from group four to group six (*i.e.* with the value of n) and that Δ decreases with increasing atomic number within a group.

An important implication of the values of $\bar{D}$(M-CH_2R) concerns metal-hydrogen bonds which are formed by both α-elimination

$$[Mo(CH_3)(CO)_2(dmpe)_2]^+ \rightarrow [Mo(H)(CO)_2(dmpe)_2]^+ \text{ (Ref.}^{64)})$$

---

[b] $\bar{E}$(Pt-Me) = 213 kJ mol^{-1}(Kharchevnikov, V. M., Rabinovich, I. B.: Trudy. Khim. Khimtechnol. **1972**, 40.

Table 8. First dissociation enthalpy D_1 (Cl_{n-1}M-Cl) kJ mol^{-1} compared with the mean dissociation enthalpy $\bar{D}$ (M-Cl) kJ mol^{-1} of MCl_n

M	n	$\bar{D}$	D_1	$\Delta = \bar{D} - D_1$
Ti	4	430	343	87
Zr	4	490	469	21
Nb	5	406	272	134
Ta	5	430	322	108
Mo	6	305	113	192
W	6	347	201	146

Values taken from Ref.[23].

and, more commonly, by β-elimination,

$[CpFe(CO)(PPh_3)(CH_2CHRR')] \rightarrow [CpFe(CO)(PPh_3)H]$ (Ref.[65])

These elimination reactions are very often thermally activated so that the M-H bonds formed in this way from σ-alkyl compounds are thermodynamically stronger than the M-C bonds from which they come. It is possible to estimate that $\bar{D}$(Ta-H) should be greater than 370 kJ mol^{-1} when produced from a Ta-CH_2CHR_2 bond, if the contrasting influences of steric and electronic effects are considered as discussed above. Kinetic measurements on the decomposition of $[CpFe(CO)(PPh_3)R]$ to form $[CpFe(CO)(PPh_3)H]$ showed an activation energy of ca. 130 kJ mol^{-1} for the process[65].

2.3. Metal-(π-Arene) Compounds, L = ArH

2.3.1. Ar = Cyclopentadienyl

For a long time the problem of defining the value of the (M-Cp) bond enthalpy contribution in MCp_2 has been aggravated by the lack of a proper determination of the enthalpy of formation of the cyclopentadienyl radical. This problem appears to have been resolved by Rabinovich who finds[66] that $\Delta H_f^\circ[C_5H_5, g] = 209.2$ kJ mol^{-1}, which is rather greater than previous estimates[67] which were in the range 100–190 kJ mol^{-1}. The enthalpies of formation of all the 3d-metallocenes have been measured recently by combustion calorimetry[68, 69]. The results are shown in Table 9, together with the value of $\bar{D}$(M-Cp) which is defined by

$$\bar{D}(\text{M-Cp}) = 1/2[\Delta H_f^\circ(\text{M, g}) + 2\Delta H_f^\circ(C_5H_5, \text{g}) - \Delta H_f^\circ(MCp_2, \text{g})]$$

The variation of $\bar{D}$(M-Cp) with ΔH_f°(M, g) is shown in Fig. 4. The deviation of Co and Ni from the linear plot is to be expected in terms of the valence m.o. electron configuration of cobaltocene and nickelocene, both of which have electrons in the anti-

Table 9. Valence m.o. electron configuration $a_{1g}^{m}\, e_{2g}^{n}\, e_{1g}^{p}$, enthalpy of sublimation, ΔH_{sub}, and standard enthalpy of formation, (ΔH_f°, g), of metallocenes. Mean bond dissociation enthalpy, $\bar{D}$(M-Cp) (Refs.[68, 69]) and (Metal-cyclopentadienyl ring) bond length, r(M-Cp). (Ref.[72]). All thermochemical results in kJ mol^{-1}

M	m	n	p	ΔH_{sub}	ΔH_f°, g	$\bar{D}$(M-Cp)	r(M-Cp) pm
V	1	2		58.6	196.6	368	228.0
Cr	1	3		63.0	241.6	284	216.9
Mn	1	2	2	72.0	273.2	213	238.3
Fe	2	4		73.2	241.4	297	206.4
Co	2	4	1	70.3	306.7	268	209.6
Ni	2	4	2	72.4	357.7	247	219.6

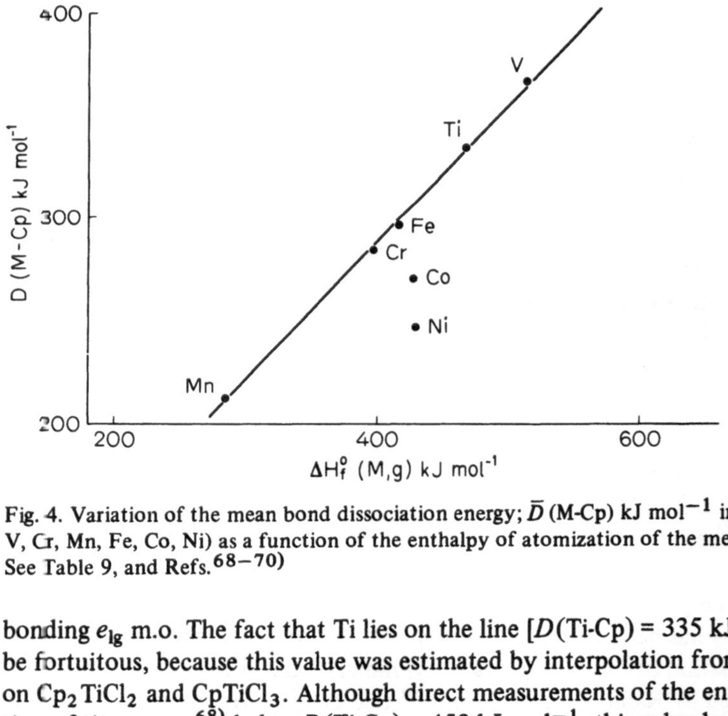

Fig. 4. Variation of the mean bond dissociation energy; $\bar{D}$(M-Cp) kJ mol^{-1} in MCp_2 (M = Ti, V, Cr, Mn, Fe, Co, Ni) as a function of the enthalpy of atomization of the metal, ΔH_f°(M, g). See Table 9, and Refs.[68–70]

bonding e_{1g} m.o. The fact that Ti lies on the line [D(Ti-Cp) = 335 kJ mol^{-1}] may be fortuitous, because this value was estimated by interpolation from measurements[70] on Cp_2TiCl_2 and $CpTiCl_3$. Although direct measurements of the enthalpy of combustion of titanocene[68] led to D(Ti-Cp) = 452 kJ mol^{-1}, this value has been ignored in favour of the lower estimate in more recent work[69]. In any case, the structure of "titanocene" is most probably[71] that shown in I so that the derivation of $\bar{D}$(Ti-Cp) from the heat of formations is not straightforward.

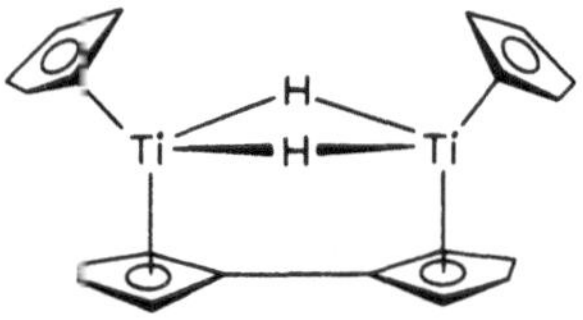

The metallocenes provide a rare opportunity to test a predicted correlation between bond strength and bond length. The structures of the compounds Cp_2M (M = V, Cr, Mn, Fe, Co, Ni) have all been studied by electron diffraction[72]. The distance between the metal atom and the plane of the cyclopentadienyl ring, r(M-Cp) is given in Table 9. If it is assumed that the loss of an electron form the a_{1g} or e_{2g} m.o. weakens the (M-Cp) bond to the same extent as the introduction of an electron into the antibonding e_{1g} m.o., then an electron imbalance can be defined[72] as the number of holes in the a_{1g} and e_{2g} m.o.s plus the number of electrons in the e_{1g} m.o. Following from this, it has been suggested that the strength of the M-Cp bond should decrease in the order M = Fe > Co > Cr > Ni > V > Mn. As Fig. 5 shows, this prediction is quite well supported by the measurements, especially with regard

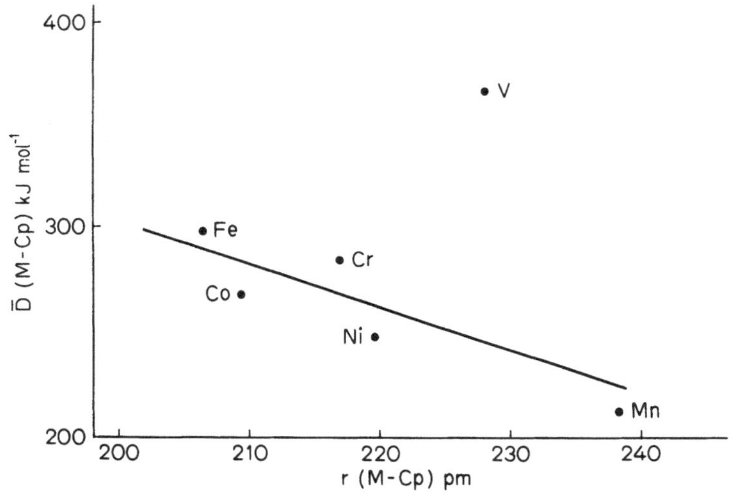

Fig. 5. Variation of the mean bond dissociation energy, $\bar{D}$(M-Cp) kJ mol^{-1}, in MCp_2 (M = V, Cr, Mn, Fe, Co, Ni) as a function of the metal-cyclopentadienyl ring plane distance, r(M-Cp) pm. See Table 9 and Refs.[68, 69, 72]

to the relative positions of chromium and nickel in that it shows that the presence of two antibonding e_{1g} electrons weakens the M-Cp bond slightly more than a hole in both the a_{1g} and the e_{2g} m.os. However, the position of vanadium is clearly anomalous and seems to suggest that the V-Cp bond in vanadocene is approximately 130 kJ mol^{-1} stronger than would be expected on the basis of this correlation. The variation of r(M-Cp) and r(M-alkyl) in a wide range of $3d$-transition metal compounds containing metals in formal oxidation states $\geqslant 2$, as a function of both atomic number and the relative energies of the metal $3d$- and $4s$ a.os, has led to the conclusion that the formation of strong bonds to the cyclopentadienyl ring depends upon the extent of M$3d$-Cpπ overlap. This overlap is expected to be less effective for the earlier members of the $3d$ series than for later members[73]. Figure 6 shows that the preferred value[62] of r(M-Cp) does not vary in a recognizable fashion with $\bar{D}$(M-Cp). This lack of correlation may simply emphasize the fact that r(M-Cp) and, certainly, $\bar{D}$(M-Cp) are sensitive to the formal oxidation state of the metal.

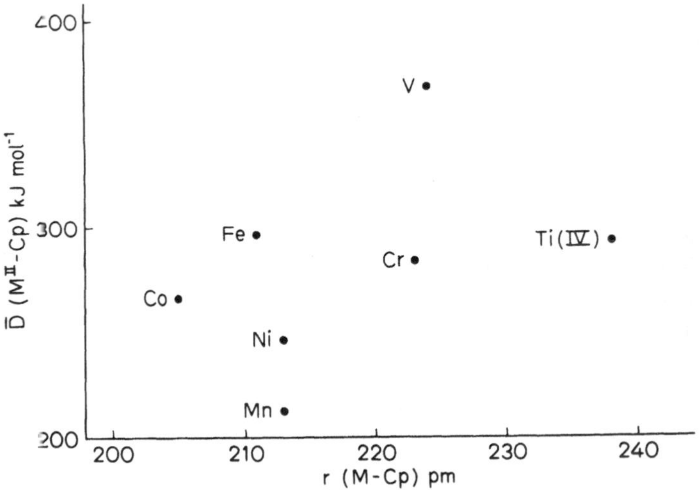

Fig. 6. Variation of the mean bond dissociation energy, $\bar{D}$ (M^{II}-Cp) kJ mol^{-1}, in MCp_2 (M = Ti, V, Cr, Mn, Fe, Co, Ni) with average metal-cyclopentadienyl ring plane distance in well determined structrues (Ref.[73]) r(M-Cp) pm. See Table 9

The value of $\bar{D}$ (M-Cp) in the metallocenes Cp_3M (M = Sc, Y, La) has been found[74] to be 305(Sc), 326(Y) and 322(La) kJ mol^{-1}. The crystal structure of Cp_3Sc shows that only two of the rings are pentahapto-bonded, the other being σ-bonded to the metal[75].

2.3.2. ArH = Benzene and Substituted Benzenes

Measurements of the heat of combustion in oxygen of bis(benzene)chromium by the conventional bomb calorimetric technique have been made on a number of occasions. The standard heat of formation ΔH_f° [$Cr(C_6H_6)_2$, c] derived from the heat of combustion was given as +89 (Ref.[76]), +209 (Ref.[77]) and +146 (Ref.[7]) kJ mol^{-1}. The discordancy reflects the difficulty of dealing with organochromium compounds by the static bomb method due to incomplete combustion of the sample. Microcalorimetric studies[8] of the thermal decomposition and reaction of bis(benzene) chromium with iodine vapour give ΔH_f° [$Cr(C_6H_6)_2$, c] = + 142 kJ mol^{-1}, from which ΔH_f° [$Cr(C_6H_6)_2$, g] = +234 kJ mol^{-1} and $\bar{D}$(Cr-C_6H_6) = 163 kJ mol^{-1}. Apart from this, very little information is available concerning $\bar{D}$ (M-ArH) bond enthalpy contributions; the values are shown in Table 10.

2.3.3. Mass Spectroscopic Determinations

The use of electron impact methods to determine bond enthapies, D(M-R), in the molecule MR_n from the relation

$$D(\text{M-R}) \leqslant \frac{1}{n} [AP(M^+) - IP(M)]$$

Table 10. Bond enthalpy contributions, D(M-ArH) kJ mol^{-1}, in bis(arene) metal compounds

M	Arene	D(M-ArH)	Ref.
V	Benzene	290	78)
Cr	Benzene	163	7, 8)
Cr	Ethylbenzene	159	79)
Cr	1,2-Diethylbenzene	167	79)
Cr	Isopropylbenzene	176	79)
Cr	1,2-Diisopropylbenzene	176	79)
Mo	Benzene	211	78)

The heats of combustion of $Cr(ArH)_2$ (ArH = 1,3,5-$C_6H_3Me_3$, 1,2,4-$C_6H_3Me_3$ (Ref.[78]), and C_6Me_6 (F. Röhrscheid, Ph. D. thesis, Munich 1965, quoted in Ref.[80]) have been reported, but lack of information about the extent of combustion makes the derivation of $\bar{D}$ (Cr-ArH) uncertain.

is accompanied by considerable problems which were outlined earlier. Where the metallocenes are concerned, the value of $\bar{D}$ (M-Cp) determined by electron impact is in the range of 30–50 kJ mol^{-1} greater than the value of this bond enthalpy obtained from combustion calorimetry. Although electron impact gives the correct progression of relative values D(Cp-V) > D(Cp-Fe) > D(Cp-Ni), it is incorrect in giving D(Cp-Mn) > D(Cp-Ni) and D(Cp-Co) > D(Cp-Fe).

While these disadvantages are severe, electron impact determinations play a useful role in suggesting the pattern of variation in bond enthalpy contributions in molecules which have not been studied by conventional thermochemical techniques. A few examples of this are shown in Table 11. Electron impact measurements also indicate that $\bar{D}$ (Ru-Cp) in ruthenocene is ca. 100 kJ mol^{-1} greater than $\bar{D}$ (Fe-Cp) in ferrocene[82].

Table 11. Bond dissociation energies determined by electron impact methods. (kJ mol^{-1})

Molecule	Bond	D(M-ArH)
Cp_2V	V-Cp	374
$(C_6H_6)_2V$	V-C_6H_6	340
$CpV(C_7H_7)$	V-C_7H_7	134
Cp_2Cr	Cr-Cp	309
$CpCrC_6H_6$	Cr-C_6H_6	280
$CpCrC_7H_7$	Cr-C_7H_7	326
$(C_6H_6)_2Cr$	Cr-C_6H_6	195
$(1,3,5\text{-}C_6H_3Me_3)_2Cr$	Cr-$C_6H_3Me_3$	257
$(C_6Me_6)_2Cr$	Cr-C_6Me_6	347
Cp_2Mn	Mn-Cp	298
$CpMn(C_6H_6)$	Mn-C_6H_6	510

Values taken from Refs.[80] and [81].

3. Substituted Metal Carbonyls

With the general pattern of bond enthalpy contributions or, more precisely, mean bond dissociation energies providing a firm background and by invoking the transferability principle (Section 1.3.), it is possible to investigate a range of important and interesting problems concerning the strengths of bonds in other organometallic compounds.

3.1. Metal Carbonyl Halides

The standard enthalpies of formation have been described[3] for a small number of carbonyl halides, $[M(CO)_m X_n]$ ($M = Mn$, $m = 5$, $X = Cl, Br$, $n = 1$; $M = Fe$, $m = 4$, $X = Br, I$, $n = 2$) and the relevant values together with the heat of disruption are given in Table 12a. If the value of $\bar{T}$(M-CO) given in Table 1 is used then the b.e.c of the M-X bond can be calculated. The enthalpies obtained in this way are less than in the gaseous metal halides at the same formal oxidation level (Table 12b). This suggests that the M-X bond may be relatively weakened by the presence of the π-acceptor carbonyl ligands attached to the metal. If the alternative assumption is made, namely that the b.e.c (M-X) remains unchanged from its value in MX_n(g), then the value of the b.e.c. (Mn-CO) falls from 99 to 88 kJ mol^{-1} and the b.e.c (Fe-CO) falls from 117 to 64 kJ mol^{-1} in $Fe(CO)_4Br_2$, and from 117 to 73 kJ mol^{-1} in $Fe(CO)_4I_2$.

Table 12a. Standard enthalpy of formation, ΔH_f(g), enthalpy of disruption, ΔH_D, and metal-halogen bond enthalpy contribution, E(M-X), in metal carbonyl halides (kJ mol^{-1})

Compound	ΔH_f° (g)	ΔH_D	E(M-X)	Ref.
$Mn(CO)_5Cl$	−950.6	799	304	3)
$Mn(CO)_5Br$	−911.3	753	258	3)
$Fe(CO)_4Br_2$	−744.3	941	236	3)
$Fe(CO)_4I_2$	−672.4	862	197	3)
$Rh_2(CO)_4Cl_2$	−710[1]	1620	100	84)

[1]) assuming ΔH_{sub} = 160 kJ mol^{-1}.

Table 12b. Bond dissociation enthalpies in gaseous metal halides (kJ mol^{-1})

D(Mn-Cl)	359
D(Mn-Br)	312
D(FeBr-Br)	439
D(Fe-Br)	247
D($RhCl_2$-Cl)	180
$\bar{D}$(Rh-Cl) in $RhCl_3$	285

Values taken from Ref.[23].

The structure of $[Rh(CO)_2Cl]_2$ in the solid state[83] shows association between the dimer molecules, so that for any one dimeric unit there will be a share in five metal-metal bonds [one r(Rh-Rh) = 312 pm; four r(Rh-Rh) = 331 pm]. The chlorine atoms bridge the short (intradimer) metal-metal bond. The b.e.cs to the heat of disruption can be written,

$$\Delta H_D = (M_1 + 4M_2) + 4\bar{T} + 4[\mu(\text{Rh-Cl})]$$

If both $\bar{M}$ and $\bar{T}$ are transferred from the binary rhodium carbonyls (Table 1) then the b.e.c of μ(Rh-Cl) is 100 kJ mol^{-1}. Alternatively, if the b.e.cs of the Rh-Rh bonds are reduced in proportion to their length (mean value in $Rh_x(CO)_y$ = 275 pm) (M_1 = 100, M_2 = 95) then the b.e.c of each bridging Rh-Cl bond increases to 120 kJ mol^{-1}.

Whichever value of this Rh-Cl bond enthalpy is adopted, it is rather lower than either $\bar{D}$(Rh-Cl) in $RhCl_3$ or $\bar{D}$($RhCl_2$-Cl), as Table 12b shows. Recent measurements of the enthalpy of dissociation of $[Rh(CO)_2Cl]_2$ in solution to form the coordinatively unsaturated species $Rh(CO)_2Cl$, gave[85] a value of 94 kJ mol^{-1}.

3.2. Metal Carbonyl Alkyls and Acyls

The heat of reaction between iodine and $CH_3M(CO)_5$ (M = Mn, Re) in the vapour phase has been measured[16] by microcalorimetry. The reaction studied is described by the equation,

$$CH_3M(CO)_5 + [(n + 1)/2]I_2 \rightarrow MI_n + CH_3 + 5CO$$

The standard enthalpies of formation of the gaseous compounds and the enthalpy of disruption derived therefrom are given in Table 13. An interesting problem arises as to how these results are to be evaluated. If the value of $\Delta H_f^\circ[M(CO)_5, g]$ derived[15, 17] from electron impact measurements on $M_2(CO)_{10}$ (M = Mn, Re) is used, then as outlined earlier this will be expected to give an *upper* limit to the value of D(M-M). It has been shown[16] that for all values of D(M-M) below specified upper limits the following relation holds

$$D(CH_3\text{-M}) = 1/2\, D(\text{M-M}) + Z$$

[where Z = 77.4 (Mn), 128.9 (Re) kJ mol^{-1}]. In this way the values of the various b.e.cs in $CH_3M(CO)_5$ in column A and B of Table 13 were derived. If instead, the values of $\bar{T}$ and $\bar{M}$ (based on the empirical relationship $\bar{M} \sim 0.68\,\bar{T}$) shown in Table 1 are used, then the b.e.cs in column C of Table 13 are obtained. If the value of $\bar{T}$ only is assumed (from Table 1) then the b.e.cs in column D of Table 13 are produced. Finally, if the value of $\bar{M}$ from Table 3 is assumed then the b.e.cs in column E of Table 13 result. The result of this is to demonstrate how the b.e.c values derived from ΔH_D are not, with the exception of B, subject to great variation no matter what assumption is made. The most important conclusion to be drawn from the

Table 13. Enthalpy of formation, ΔH_f(g) and disruption, ΔH_D (kJ mol^{-1}) of alkyl and acyl derivatives of manganese and rhenium carbonyls

Compound	ΔH_f° (g)	ΔH_D	Ref.
$CH_3Mn(CO)_5$	−730.5	601	16)
$CH_3Re(CO)_5$	−763.1	1123	16)
$PhCOMn(CO)_5$	−753	590	72)

Bond enthalpy contributions (kJ mol^{-1}) in $CH_3M(CO)_5$

	Bond	A[1]	B[2]	C	D	E
M = Mn						
	$Mn-CH_3$	117	129	111	106	113
	Mn-CO	97	95	100[3]	100[3]	98
	Mn-Mn	79	104	67[3]	59	71[4]
M = Re						
	$Re-CH_3$		223	192	188	193
	Re-CO		181	187[3]	187[3]	186
	Re-Re		187	128[3]	118	129[4]

[1] Based on Ref.[15].
[2] Based on Ref.[17].
[3] Value taken from Table 1.
[4] Value taken from Table 3.

measurements is that the ($M-CH_3$) bond is comparable to or stronger than the M-CO bond strength. It is worth emphasizing that there is no necessary identity between the b.e.c and the corresponding bond dissociation energy, because the former does not take account of reorganization energy released in the disruption. Nonetheless, the finding that M-alkyl ⩾ M-CO is consistent with the earlier discussion of Sections 2.1. and 2.2.

The benzoyl derivative, $PhCOMn(CO)_5$ and related acyl derivatives have intrinsic interest because they can be prepared by alkyl group migration to coordinated CO, a formal internal nucleophilic attack. This reaction, which is often referred to as carbonyl insertion[86], is reversible in certain instances to give the metal alkyl. The enthalpy of disruption for $PhCOMn(CO)_5$ (Table 13) can be divided up to give the b.e.c of the Mn-COPh bond; if the values of $\overline{T}$ (Mn-CO) shown in Table 13 are used, then the Mn-COPh enthalpy lies in the range 105 ± 10 kJ mol^{-1}, the lower value being preferred as outlined earlier.

The overall enthalpy change of the insertion process contains contributions from four bonds (M-CO, M-COR, M-R and CO-R). As there is no significant difference between E(Mn-R) and E(Mn-COR) then, at least in the case of manganese and hydrocarbon groups, R, the dominant factor will be the difference between $\overline{T}$ (Mn-CO) and E(R-COX); [for $R = CH_3, E = 339$ kJ mol^{-1} (X = H), 370 kJ mol^{-1} (X = Cl) (Ref.[23])] which suggests that the insertion reaction is thermodynamically favoured with respect to decarbonylation. Kinetic studies of the carbonyl insertion reaction in solution have shown[87] that the enthalpy of activation is 62 kJ mol^{-1} for insertion, and 118 kJ mol^{-1} for decarbonylation in the reaction of $CH_3Mn(CO)_5$ with carbon monoxide.

Differential scanning calorimetry has been used[88] to measure the enthalpy change, ΔH_0 for the exothermic decarbonylation reaction

$$[Ir(COR)Cl_2(PPh_3)_2c] \rightarrow [Ir(CO)(R)Cl_2(PPh_3)_2, c]$$

in the solid state. The value of ΔH_0 is sensitive to the nature of the group R, increasing linearly with the number of fluorine atoms from R = CF_3 (ΔH_0 = = −83 kJ mol^{-1}) to R = CH_3 [ΔH_0 = −27 kJ mol^{-1} (extrapolated)]. On the basis of these measurements, it has been estimated that the (Ir-CF_3) bond may be at least (57 ± 9) kJ mol^{-1} stronger than the (Ir-CH_3) bond. This result provides an explanation for the fact that carbon monoxide insertion into a (perfluoroalkyl-metal) bond has not been observed, whereas perfluoroacyl-metal complexes are more readily decarbonylated than their hydrocarbon analogues[86]. An estimate of the enthalpy of activation for carbon monoxide insertion into an (Ir-CF_3) bond gives a value of +178 kJ mol^{-1} so that the reaction will be slow[88]. A variety of 4-substituted benzyl groups (R = 4–X · $C_6H_4CH_2$) all give values of ΔH_0 in the range (−15 ± 1) kJ mol^{-1}, for the decarbonylation reaction. The influence which phase effects and changes in coordination number and geometry occurring in this reaction will have on the enthalpy are unknown, but they are assumed to be small and mutually cancelling.

3.3. Complexes of Nitrogen-Donor Ligands

Differential scanning calorimetry has been used to measure[89, 90] the enthalpy of reaction for the displacement of a nitrogen donor ligand, L, from tungsten complexes $[W(CO)_{6-n}L_n]$ by carbon monoxide under isobaric conditions. The reaction is described by the equation

$$[W(CO)_{6-n}L_n] + nCO \rightarrow W(CO)_6 + nL$$

The results of these investigations are shown in Table 14 where E(W-N) = $\frac{1}{n}$

$$E(W\text{-}N) = \frac{1}{n}[\Delta H_D - (6-n)\bar{T}],$$

Table 14. Standard enthalpies of sublimation, formation and disruption and bond enthalpy contributions, E(W-N) kJ mol^{-1}, for N-donor complexes of tungsten $[W(CO)_{6-n}L_n]$

L	n	ΔH_{sub}	$\Delta H_f^\circ(g)$	ΔH_D[1]	E(W-N)	Ref.
4-Methylpyridine	1	30.5	−776	1175	285	89)
Pyridine	1	30.5	−679	1135	245	89)
Piperidine	1	53.1	−878	1128	238	89)
CH_3CN	1	48.1	−703	1078	188	90)
CH_3CN, *cis*	2	57.3	−498	1057	172	90)
CH_3CN, *fac*	3	43.9	−315	1058	176	90)

[1]) Using ΔH_f° (CH_3CN, g) = 73.7 kJ mol^{-1} (G. Pilcher, unpublished work).

using the value of $\bar{T}$ in Table 1 (178 kJ mol^{-1}). When it is remembered that the heat of sublimation of $W(CO)_6$ is 74 kJ mol^{-1}, the heats of sublimation for all of the compounds in Table 14 are surprisingly low. An alternative method of evaluating E(W-N) in the acetonitrile complexes, $[W(CO)_{6-n}(MeCN)_n]$ (n = 1,2,3) uses an approximation based on electron impact data to obtain $\Delta H_f[W(CO)_{6-n},g]$. This method[90] also gives an almost constant b.e.c for (W-NCMe) which is in the range 174–182 kJ mol^{-1} for all three compounds. The difference in E(W-N) among the various ligands is very large. It is suggested[89], that the variations can be explained in terms of the σ- and π-bonding characteristics, so that the increase in D(W-N) from pyridine to 4-methylpyridine (40 kJ mol^{-1}) is taken to show the presence of a stronger (N $\rightarrow$ W) σ interaction; and yet, notwithstanding the difference in pK_a-values (5.17 and 11.2) and gas phase proton affinities[91] (940 and 961 kJ mol^{-1}) pyridine (which can function as a π-acceptor) and piperidine (which functions exclusively as a σ-donor) have approximately the same b.e.cs with respect to tungsten. The lower basicity of MeCN [PA 787 kJ mol^{-1} (Ref.[78])], may perhaps be reflected in the lower value of E(W-N).

Measurements[92] of the enthalpy of thermal decomposition and of iodination of the complexes *fac*- $[Mo(CO)_3L_3]$ (L = MeCN, pyridine) lead to $\bar{E}$ (Mo-NCMe) ~ 130 kJ mol^{-1} and $E(Mo\text{-}NC_5H_5)$ ~ 150 kJ mol^{-1}. These results indicate that whereas $\bar{E}$ (Mo-NCMe) is slightly less than $\bar{T}$ (Mo-CO), the b.e.c of the better π-acceptor ligand, $\bar{E}$ $(Mo\text{-}NC_5H_5)$ is very similar to $\bar{T}$ (Mo-CO). In such discussions of the influence of the donor/acceptor characteristics of ligands upon the b.e.c of (M-L), it is useful to recall that[47] $\bar{D}$(M-CO) = $\bar{D}$$(M\text{-}PF_3)$ (M = Cr, Ni) (see Section 2.1.) so that the range of b.e.cs may not be very great in spite of a profound change in the character of L. This conclusion is supported by the results of ion cyclotron resonance measurements[93] on a variety of $(CpNiL)^+$ compounds (see Section 3.8. and Table 23) and a calorimetric study[94] of the reaction between various tertiary phosphines, phosphites and Bu^{t}NC and di(1,5-cyclooctadiene)nickel (see Section 4. and Table 24).

3.4. π-Arene Derivatives

3.4.1. Arene Tricarbonyl Complexes of Group Six Metals

Microcalorimetric measurements of the heat of thermal decomposition and reaction with iodine of various $(ArH)M(CO)_3$ compounds are collected in Table 15. The b.e.c of the (M-ArH) bond, E(ArH-M) is derived from ΔH_D assuming the $\bar{T}$ (Cr-CO) is transferred from $Cr(CO)_6$ (Table 1). These b.e.cs increase with increasing methyl substitution at the benzene ring. That the strength of the ArH-M bond should increase in this manner is consistent with the bulk of evidence from spectroscopic measurements[95]. Likewise, Table 15 shows that introduction of an electron-withdrawing group to the benzene ring, as in chlorobenzene, reduces E(ArH-M). The use of $[(\eta^6\text{-cycloheptatriene})Cr(CO)_3]$ as a precursor in the synthesis of other $L_3M(CO)_3$ complexes is well known. The value of E(ArH-M) in this compound indicates that thermochemically, one may expect a non-aromatic 6-electron donor ligand to be

Table 15. Enthalpy of sublimation, formation and disruption (kJ mol^{-1}) for [π-arene $M(CO)_3$] compounds and arene-M bond enthalpy contributions E(ArH-M) kJ mol^{-1}

M	Arene	ΔH_{sub}	ΔH_f° (g)	ΔH_D	E(ArH-M)	Ref.
Cr	Benzene	91.2	−352	502	178	5, 8)
Cr	Toluene	94.6	−382	498	174	5)
Cr	Mesitylene	108.4	−463	512	191	84)
Cr	Hexamethylbenzene	123.4	−548	526	205	84)
Cr	Chlorobenzene	102.5	−364	481	157	5)
Cr	Cycloheptatriene	88	−222	471	150	84)
Mo	Mesitylene	109	−424	735	279	84)
Mo	Cycloheptatriene	88	−209	720	251	92)
W	Mesitylene	111	−367	870	334	84)
W	Cycloheptatriene	92	−144	847	311	92)

[1]) Estimated value.

displaced rather more easily than the other aromatic 6-electron donor ligands (*e.g.* benzene) from chromium.

If the arene ligand is kept the same and the metal changed, Table 15 shows that D(M-mesitylene) increases with increasing atomic number. In Fig. 7 the variation

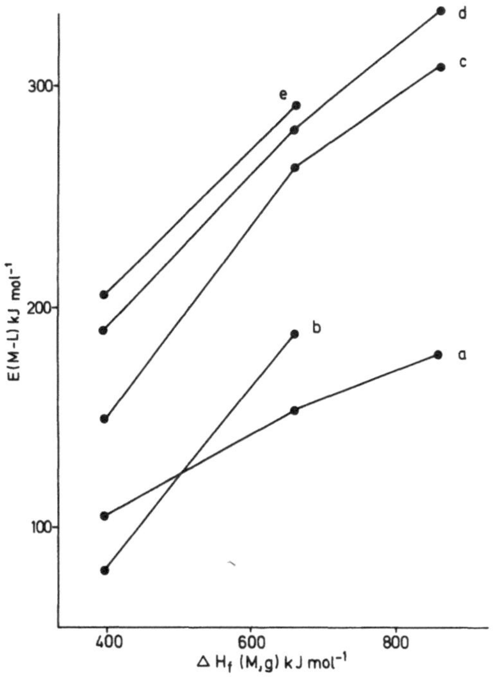

Fig. 7. Variation of the bond enthalpy contribution, E(M-L) kJ mol^{-1} (M = Cr, Mo, W) as a function of the enthalpy of atomization of the metal, ΔH_f° (M, g) kJ mol^{-1}. a) $\bar{T}$ (M-CO); b) E(M-norbornadiene); c) E(M-cycloheptatriene); d) E(M-mesitylene); e) E(M-hexamethylbenzene)

of both E(ArH-M) and $\overline{T}$(M-CO) with ΔH_f°(M, g) are compared, and it would appear that the former increases more rapidly than the latter.

Electron impact studies of $ArHM(CO)_3$[96] and $(AH)_2M$ compounds[81] are consistent with the pattern of enthalpies shown in Table 15, but as may be expected, the differences are overestimated. Thus, $D(Cr\text{-}C_6Me_6)$ is indicated to be greater than $D(Cr\text{-}C_6H_6)$ by ca. 150 kJ mol^{-1} by electron impact measurements[80], three times the difference found by calorimetry. It is perhaps pertinent to note at this point that, even if the principle of transferability (Section 1.3.) is not accepted, so that E(ArH-M) = 0.25 ΔH_D as a result of making E(ArH-M) = D(M-CO) for example, then the same pattern of relative values of E(ArH-M) shown in Table 15 is obtained but with $E(Cr\text{-}C_6H_6)$ = 125 kJ mol^{-1} and $E(Cr\text{-}C_6Me_6)$ = 138 kJ mol^{-1}. There are a number of reasons, based on the chemical and spectroscopic properties of these compounds, why it would appear unlikely that E(ArH-M) should take the same value as D(M-CO).

When the same microcalorimetric methods are applied to hexaalkylborazole complexes, $[(R_3B_3N_3R_3)Cr(CO)_3]$ (R = Me, Et), it is possible to show[84] that, for both complexes, the b.e.c of the (Cr-borazole) bond should lie in the range 70–140 kJ mol^{-1}, with the true value probably in the range (105 ± 12) kJ mol^{-1}. This value of $E(Cr\text{-}R_6B_3N_3)$ is much less than $E(Cr\text{-}C_6Me_6)$, with which an isoelectronic comparison can be made. The crystal structure of $[(Et_6B_3N_3)Cr(CO)_3]$ shows[97] that the borazole ring is bound to chromium through the three nitrogen atoms. Attempts to measure D(Cr-N) in other nitrogen donor complexes of chromium(O) indicate[84] that the b.e.c (Cr-pyridine) in *cis*-$[Cr(pyr)_2(CO)_4]$ is in the range 90–105 kJ mol^{-1} (see also Section 3.3.). If allowance is made for some (B-N) π-bonding in the complex and for deformation of the ring[97] on complexation, the (Cr-borazole) b.e.c is consistent with that for (Cr-pyridine), and reflects the low thermal stability of the compound and the great ease with which the borazole ring can be displaced[98].

3.4.2. $[(\pi\text{-Arene})Co_4(CO)_9]$ Complexes

Compounds of the type $[(\pi\text{-arene})Co_4(CO)_9]$ have been prepared by Pauson[99]. They are formally isoelectronic with the $(ArH)Cr(CO)_3$ series, and are derived from $Co_4(CO)_{12}$. The thermal decomposition of three representatives of the series has been studied by microcalorimetry[84] and the results are shown in Table 16. Once again heats of sublimation have had to be estimated by comparison with the chromium analogues. The enthalpy disruption can be divided by taking $\overline{T}$ = 134 kJ mol^{-1} (Table 1) so that the b.e.c of the $[Co_4(CO)_9]$ fragment in $Co_4(CO)_{12}$ is 1722 kJ mol^{-1}. The (ArHCo) bond enthalpy contribution is then obtained in the usual way; the results are shown in Table 16. It is clear that as in the chromium series, the b.e.c (ArH-Co) increases along the series benzene < mesitylene < hexamethylbenzene.

3.4.3. π-Cyclopentadienyl Metal Carbonyl Compounds

Probably on account of the lack of reliable D(M-Cp) values until recently[69], only one compound in this class has been investigated. Combustion of $CpMn(CO)_3$ gave[100]

Table 16. Enthalpy of formation, ΔH_f° (c) and disruption, ΔH_D and bond enthalpy contribution, E (Co-L) in $LCo_4(CO)_9$. All values in kJ mol^{-1} (Ref.[84])

L	ΔH_f° (c)	ΔH_D	E(Co-L)
$(CO)_3$	−1845	2130	402
Benzene	−1313	1984	270
Mesitylene	−1444	1998	284
Hexamethylbenzene	−1555	2023	310

$\Delta H_f^\circ [CpMn(CO)_3, c] = -525.1 \pm 4$ kJ mol^{-1}. We may estimate $\Delta H_{sub} = 90$ kJ mol by comparison with $[C_6H_6Cr(CO)_3]$ and Cp_2Fe, so that $\Delta H_f^\circ [CpMn(CO)_3, g] = -435$ kJ mol^{-1}. The enthalpy of disruption, $\Delta H_D = 597$ kJ mol^{-1} using values of ΔH_f° (g) given earlier. If the value of $\bar{T}$ (Mn-CO) is taken from Table 1, one obtains the b.e.c (Cp-Mn) = 297 kJ mol^{-1} which is rather greater than $\bar{D}$ (Cp-Mn) in Cp_2Mn (215 kJ mol^{-1}). A number of explanations can be offered for this increase including the change (decrease) in formal oxidation state, the decrease in Mn-ring bond distance (216.5 pm) and the absence of electron imbalance (Section 2.3.2.) in this 18-electron compound. Moreover this value of (Cp-Mn), is the same as (Cp-Fe) in ferrocene[58], with which $CpMn(CO)_3$ is isoelectronic.

3.5. Olefin Complexes of Metal Carbonyls

The b.e.cs of metal-olefin bonds are almost unknown, although olefin complexes[101] of metals constitute one of the most important areas within the field of organometallic chemistry. This crucial lack of quantitative information has begun to diminish in the past two years, largely through the work of Partenheimer[102], Tolman[103] and others[104] on olefin compounds of the group eight metals. Where olefin complexes of metal carbonyls are concerned, microcalorimetric measurements of the thermal decomposition and heat of reaction with iodine of a few $[M(CO)_n(olefin)_q]$ compounds have been reported[5]; the results are shown in Table 17. The average value of the (diene-Fe) b.e.c is approximately 184 kJ mol^{-1}, which is slightly less than double that of (ethylene-Fe) in $[Fe(C_2H_4)(CO)_4]$. The conjugated dienes thus appear to be bound to iron as if the two double bonds were largely independent of one another. The $(C_2H_4$-Fe) b.e.c is smaller than $\bar{T}$ (Fe-CO) and compares in this respect with rhodium[105] and nickel[103] as shown in Table 18.

Measurements of the enthalpy of thermal decomposition of five co-ordinate perfluoro-olefin and -acetylene-iridium(I) complexes according to the equation

$$[IrX(CO)(PPh_3)_2(L), c] \rightarrow [IrX(CO)(PPh_3)_2, c] + L, g$$

have been made by differential scanning calorimetry[104]. The measured enthalpy changes, ΔH_{dec}, identify with the (Ir-L) b.e.c only insofar as the heats of sublimation of the solid reactants and solid products are assumed to be identical. The results obtained are shown in Table 19. The influence of the conformation of the

Table 17. Enthalpy of sublimation/vaporization, $\Delta H_{sub/vac}$, enthalpy of disruption, ΔH_D, and olefin-iron bond enthalpy contribution for $[Fe(CO)_n(olefin)_q]$ compounds. All values are in kJ mol^{-1}

Olefin	q	n	$\Delta H_{sub/vap}$	ΔH_D	$\bar{E}$ (Fe-L)
Ethylene	1	4	41.8	567	99
Buta-1,3-diene	1	3	48.9	554	203
Buta-1,3-diene	2	1	76.1	482	183
Methyl penta-2,4-dienoate	1	3	72.4	515	164
Methyl penta-2,4-dienoate	2	1	118.0	489	186
Cyclohexa-1,3-diene	2	1	95.0	500	192

$\bar{D}$ (Ni-diene) in $[Ni(1,5\text{-cyclooctadiene})_2]$ has been estimated to be 206 kJ mol^{-1} (Ref.[94]).

Table 18. Comparison of $(M\text{-}C_2H_4)$ and (M-CO) bond enthalpy contributions for M = Fe, Ni and Rh

M	$\bar{T}$ (M-CO)	$E(M\text{-}C_2H_4)$	Ref.	$M\text{-}CO/M\text{-}C_2H_4$
Fe	117	99	5)	1.18
Ni	147	138	103)	1.07
Rh	166	130	105)	1.28

Table 19. Enthalpy of thermal decomposition, ΔH_{dec} kJ mol^{-1}, of iridium(I) complexes. $[IrX(CO)(PPh_3)_2(L)]$. (Ref.[104])

	$L = [CF_3C{\equiv}CCF_3]$	$L = [CF_2{=}CF_2]$
X	ΔH_{dec}	ΔH_{dec}
F	99	79
Cl	96	67
Br	79	41
I	82	57

$[IrX(CO)(PPh_3)_2]$ moiety in both the reactants and products upon ΔH_{dec} is unknown. The variation in ΔH_{dec} as a function of X (F > Cl ⩾ Br < I) has been used as an index of the donor/acceptor character of the (Ir-L) bond, but this will be influenced by the contribution of metallocyclic structures to the correct description of this bond.

3.6 $[(\text{Halo-Methylidyne})Co_3(CO)_9]$ Compounds

Reaction calorimetry in solution has been used[19] to measure the heat of reaction between halogens and the compounds $[Co_3(CX)(CO)_9]$ (X = Cl, Br). The enthalpies

Table 20. Enthalpy of sublimation, formation and disruption of halomethylidynetricobaltennea-carbonyl compounds $[Co_3(CX)(CO)_9]$ (kJ mol^{-1}) (Ref.[19])

X	ΔH_{sub}	ΔH_f° (g)	ΔH_D	$\bar{E}$ (Co-CX)
Cl	117.6	−1069	2191	134
Br	99.6	−1090	2179	155

of sublimation of these compounds have been determined directly by effusion manometry. The partition of the enthalpy of disruption for these compounds (Table 20) is problematical, particularly in relation to the value to be accepted for the b.e.c (C-X). One solution[19] to this problem has been to equate $[3E(\text{Co-C}) + E(\text{C-X})]$ with $[\Delta H_D - (3\bar{M} + 9\bar{T})]$ and by taking $\bar{D}$(C-X) from CX_4 as the value of E(C-X), to obtain the (Co-C) b.e.c in the range 145 ± 25 kJ mol^{-1}. The average value of the metal-carbon bond energy is comparable with $\bar{T}$(Co-CO) (134 kJ mol^{-1}) as was found in the case of $CH_3Mn(CO)_5$ and $PhCOMn(CO)_5$ (Section 3.2.).

3.7 Metal-Hydrogen Bond Enthalpy Contributions. Metal Carbonyl Hydrides

There is almost no reliable information about the strength of metal-hydrogen bonds in organometallic compounds. Earlier (Section 2.2.) it was established that M-H bonds formed by (spontaneous) β-elimination from metal alkyls should be stronger that their precursor, but this is an unsatisfactory and imprecise position. The dissociation

Table 21a. Dissociation energies of diatomic transition metal hydrides, D(M-H) kJ mol^{-1}

Cr-H	276 ± 40				
Mn-H	230 ± 30				
Ni-H	285 ± 12			Pt-H	335 ± 8
Cu-H	276 ± 8	Ag-H	222 ± 12	Au-H	310 ± 25

Values are taken from Ref.[23].

Table 21b. Initial enthalpy of adsorption of hydrogen on evaporated metal films, Q(M-H) (kJ mol^{-1})

				Ta	188
Cr	188	Mo	167	W	188
Mn	71				
Fe	134				
		Rh	109		
Ni	126	Pd	109		

Data taken from Hayward, D. O., in: Chemisorption and reactions on metallic films. Anderson, J. R., (ed.) London: Academic Press 1971, p. 226.

energies of a few diatomic MH molecules have been measured and these are summarized in Table 21a. Following the discussion of b.e.cs of (M-halogen) bonds in metal carbonyl halides, it appears reasonable to presume that the b.e.c (M-H) in a metal carbonyl hydride such as $HMn(CO)_5$ will be less than the value given in Table 21. An estimate of the Mn-H bond strength in $HMn(CO)_5$ based on electron impact measurements gave[106)] a value of (29 ± 50) kJ mol^{-1}. Application of the same method to $HCo(CO)_4$ give[107)] a value of (17 ± 63) kJ mol^{-1} for the bond enthalpy contribution of the Co-H bond. The standard enthalpy of formation $\Delta H_f^\circ [HCo(CO)_4, g] = -569$ kJ mol^{-1} has been estimated[108)] from equilibrium data. The enthalpy of disruption to H(g), Co(g) and 4CO(g), $\Delta H_D = 773$ kJ mol^{-1}, so that taking $\bar{T}$(Co-CO) from Table 1, produces a value for the b.e.c (Co-H) = 249 kJ mol^{-1}. A more recent study by infrared spectroscopy[115)] of the equilibrium between hydrogen and $Co_2(CO)_8$ in the range 350–430K and 8–16 MPa leads to $\Delta H_f^\circ [HCo(CO)_4, g] = -611$ kJ mol^{-1} and E(Co-H) = 291 kJ mol^{-1}.

The use of ion cyclotron resonance spectroscopy to measure the proton affinity of a molecule in the gas phase is now well established (for example, Ref.[91)]). The application of the technique to transition metal organometallic compounds is a more recent development and some results are shown in Table 22. In all molecules studied so far it is generally observed that the dissociation energy of a cationic

Table 22. Proton affinity and ionization potential of some transition metal compounds (ML_n) and dissociation energy of the cation $D[(L_nM\text{-}H)^+]$. All values in kJ mol^{-1}

ML_n	*PA*	*IP*	$D[(L_nM\text{-}H)^+]$	Ref.
$Fe(CO)_5$	854	770	310 ± 12	109)
$FeCp_2$	891	656	234 ± 25	110)
$NiCp_2$	916	598	202 ± 12	111)

hydride is greater than that of an isoelectronic neutral hydride, for example $D[(PH_3\text{-}H)^+] > D[(SiH_3\text{-}H)]$. On this basis it is to be expected[109)] that $E[(CO_5Mn\text{-}H]$ will be less than 310 kJ mol^{-1}. An estimate, based on the b.e.c of Mn-CH_3 in $CH_3Mn(CO)_3$ would suggest that $E[(CO)_5Mn\text{-}H]$ may be in the range 200 ± 20 kJ mol^{-1}. While there is much room for improvement of these values, it would appear unlikely that the (M-H) b.e.c in metal carbonyl hydrides can be as low as the value derived from electron impact measurements of appearance and ionization potentials would suggest.

The enthalpy of adsorption of hydrogen onto evaporated metal films, Q(M-H) has been measured for a few metals (Table 21b). Attempts[112)] have been made to calculate $Q(M\text{-}H) = 2E(M\text{-}H) - D(H\text{-}H)$, where E(M-H), the enthalpy of the metal-hydrogen bond, is given approximately by the equation due to Pauling.

$$E(M\text{-}H) = 1/2\ \{E(M\text{-}M) + D(H\text{-}H)\} + 23.06\ (\chi_M - \chi_H)^2$$

in which E(M-M) is the bond energy of a *single* metal-metal bond, which may be the value of $\bar{M}$ in Table 3, D(H-H) is the dissociation enthalpy of hydrogen (432.07 kJ mol^{-1}), and χ_M and χ_H are the electronegativity of M and H respectively.

3.8 Ion Cyclotron Resonance Spectroscopy

The use of this technique was briefly mentioned in connection with measurements of M-H bond enthalpies (Section 3.7.). A glimpse of the versatility and potential of this technique is provided by measurements[93)] of the binding energies of donor ligands to the cyclopentadienyl nickel cation, $CpNi^+$. A selection of some of the nickel-ligand b.e.cs obtained in this way is shown in Table 23. The ligands, L, were used to displace NO from CpNiNO in the gas phase. Weakly basic ligands which do not displace NO under these conditions include CO, H_2O and H_2S. In general the measurements showed that $D[(CpNi\text{-}L)^+]$ increases,

Table 23. Selected nickel-ligand bond enthalpy contributions, $D[(CpNi\text{-}L)^+]$ kJ mol^{-1} determined by ion cyclotron resonance (Ref.[93)])

Ligand, L	$D[(CpNi\text{-}L)^+]$	Ligand, L	$D[(CpNi\text{-}L)^+]$
NH_3	219	Me_2O	197
NMe_3	236	Me_2S	214
PH_3	191	Me_3COH	213
PMe_3	241	Me_3CCHO	208
$AsMe_3$	240	MeOH	191
HCN	200	MeSH	197
MeCN	223	MeCHO	197
NO	192	MeNC	241
		CO	<190

a) with alkylation of the donor atom,

b) with increasing methyl substitution of carbon atoms both proximal (α-) and distant from the donor atom, and

c) on passing from a first row to a second row *n*-donor ligand.

The relative binding enthalpies, $\delta D[(CpNi\text{-}L)^+]$ show a good linear correlation with the relative proton affinities of $\delta D[(L\text{-}H)^+]$ of the ligands, except that ligands having π- bonding ability (NO, MeCN, MeNC) are more strongly bound to nickel than would be expected from their proton affinity alone.

Following the discussion in previous sections, it is reasonable to suggest that, while the absolute magnitude of these b.e.cs will change when the same ligands are attached to other metals in neutral compounds, their relative magnitude will not change greatly. In this context it is noteworthy that the range of enthalpy *difference* spanned by the various ligands mentioned in Table 23 is less than 55 kJ mol^{-1}.

3.9 Metal-Metal Bond Enthalpy Contributions

The determination of this quantity, which is of interest in many different types of metal carbonyl compound, is a contentious matter and relatively few measurements have been made. Mass spectrometric determinations of metal-metal bond enthalpies have been made[15, 17, 117, 120], for a small number of binuclear metal carbonyls, but the assumptions necessary in some cases[17] (*e.g.* $MnRe(CO)_{10}$) may lead to a substantially overestimated result. Kinetic measurements of substitution and thermal decomposition reactions of these compounds in solution can be used to derive activation enthalpies for metal-metal bond homolysis. Where a comparison can be made between the two methods, the discrepancy has been ascribed either to the excitation energies of the radicals produced by electron impact or to the activation enthalpy for radical recombination in solution. The results of the two approaches are presented in Table 24. Other estimates of metal-metal b.e.cs have been considered in Section 2.1 (see also Tables 1,2 and 3).

Table 24. Metal-metal bond enthalpies (kJ mol^{-1}) in binuclear metal carbonyl complexes determined by electron impact and reaction kinetic methods

Compound	D(M-M) by Electron Impact	Ref.	$\Delta H^{\neq}$ for M-M Bond homolysis	Ref.
$Mn_2(CO)_{10}$	79	15)	151	116)
	88	117)		
	105	17)		
$Tc_2(CO)_{10}$	177	17)	160	118)
$Re_2(CO)_{10}$	187	17)	162	119)
$MnRe(CO)_{10}$	210	17)	163	118)
$Co_2(CO)_8$	54	117)		
$[CpW(CO)_3]_2$	234	110)		
$[CpFe(CO)_2]_2$			96	121)
$[Co(CO)_3(PBu_3)]_2$			110	122)

4. Conclusions

The foregoing account has shown how the pattern of b.e.cs in organo-transition metal carbonyl compounds has developed to data. Although the b.e.cs of a number of the principal classes of metal-ligand bond have been investigated, there remain many classes about which little or nothing is known with precision. Examples of such classes would include metal-alkyne, metal-carbene, metal-enyl and metal-hydride bond enthalpies, in particular. In the wider context, very little is known about the effects of ring size, or about such matters as double bond conjugation, but there are indications that the influence of these factors on the b.e.c will either be undetectable or rather small.

There is a lack of information about b.e.cs involving 4*d*- and 5*d* transition metals. The detailed variation of b.e.cs with the formal oxidation state of the metal remains to be defined beyond the generalization that there is an approximate inverse relation between the two. However, the comparison of $\bar{D}$(W-CO) with $\bar{D}$ ($W-CH_3$) indicates that the bond enthalpy/oxidation state relationship is complex.

Relatively little is yet known about the influence of electronic effects by substituents on b.e.cs, but such data as there are tend to show that this influence, while not individually great, is cumulative, as for example in π-arene complexes, $(ArH)Cr(CO)_3$ (Table 15). Even less is known about the influence of steric effects, but studies on the formation of phosphine complexes of nickel have shown[94)] that an increase in steric strain reduces the value of $\bar{D}$(Ni-P) which, however, is not strongly dependent on the electron donor/acceptor properties of the phosphorus or other ligand attached to nickel (Table 25). Once again (cf. Table 23, for data on cationic nickel compounds) the range of enthalpy difference spanned by the ligands in Table 25 is small (less than 25 kJ mol^{-1}), although the ligands are very different from one another.

Table 25. Selected nickel-ligand bond enthalpy contributions, $\bar{D}$(Ni-L), kJ mol^{-1} in NiL_n as a function of ligand cone angle, θ, (Ref.[113)]) and strain energy, Q, kJ mol^{-1}

L	n	θ (deg)	Q[1)]	$\bar{D}$(Ni-L)	Ref.
PF_3	4	104		147	47)
$P(OMe)_3$	4	107	0	157	94)
$P(OPr^i)_3$	4	128	29	150	94)
$PPhMe_2$	4	122	54	143	94)
PMe_3	4	118	63	141	94)
PPh_2Me	4	136	79	137	94)
PPh_3	3	145	>138	149	94)
CO	4			147	20)
Bu^tNC	4			143	94)

[1)] Strain energy relative to $P(OMe)_3$.

The principal assumption of transferability (Section 1.3.) seems to be acceptable on the basis of the data accumulated so far, but clearly requires further testing where organic ligands are concerned. No alternative assumption suggests itself in relation to M-CO bond enthalpies, but the resulting values of $\bar{T}$ and $\bar{M}$ could be improved when measurements are made on the more exotic polynuclear metal carbonyls that have been prepared in the last few years.

Where it is possible to compare b.e.cs of the same ligand to different metals of the same transition series, as in the case of the 3*d*-metals chromium, manganese, iron, cobalt and nickel, the most striking features are the narrow range (ca. 200 kJ) within which all the presently known b.e.cs of organic ligands fall, and the even smaller variation in E(M-L) or $\bar{D}$(M-L) for a particular ligand L as the metal changes. These facts are summarized in Table 26, together with a selection of other metal-ligand bond enthalpies for comparison.

It is to be hoped that measurements will be made in the near future which will put more substantial flesh on the skeleton of known bond enthalpy contributions in organo-transition metal compounds, so that a better understanding of the energetics of reactions such as olefin disproportionation (metathesis) and hydroformylation may be achieved.

Table 26. Selected bond enthalpy contributions, $\bar{D}$(M-L) and E(M-L) kJ mol^{-1}, in organometallic compounds of chromium, manganese, iron cobalt and nickel and related compounds (An asterisk (*) denotes the average value in a series)

Ligand, L	M = Cr	Mn	Fe	Co	Ni	
CO ($\bar{T}$)	108	100	117	136	147	[1]
M ($\bar{M}$)		67	80	86		[1]
M	151	42	100	167	258	[2]
Cl	385	393	393	376	368	[3]
Cp	284	213	219	268	247	[4]
ArH	168*					[5]
ArH	185*			275*		[6]
alkyl		110		145*		[7]
C_2H_4			99		138	[8]
diene			186*		206	[8]
RNC					143	[9]
PR_3					146*	[9]

[1] In $M_m(CO)_n$.
[2] In M_2, Table 2.
[3] In MCl_2, Ref.[23].
[4] In Cp_2M, Table 9.
[5] In $(ArH)_2M$, Table 10.
[6] In (ArH) $M_m(CO)_n$, Table 15 (Cr) and 16 (Co).
[7] In (alkyl) $M_m(CO)_n$, Table 13 (Mn) and 20 (Co).
[8] Tables 17 and 18.
[9] In NiL_4, Table 25.

Acknowledgements. It is a particular pleasure to record my indebtedness to Dr. H. A. Skinner both for his comments on a draft of this review and for much else besides. Thanks are due to the Science Research Council for its support, to the various people at Manchester who, during the past five years, have made some of the measurements reported here, and to those in other places who have generously made materials available for study.

5. References

[1] Skinner, H. A.: Combustion calorimetry of organometallic compounds. In: experimental thermochemistry. (editor, 2nd edit. in course of publication); Advances in Organometallic Chemistry *2*, 49 (1964)
[2] Pilcher, G.: Thermochemistry of organometallic compounds containing metal-carbon linkages. MTP Review of Science. Series 2. Physical chemistry, Vol. 10. Chapter 2. 1975, p. 45; Cox, J. D., Pilcher, G.: Thermochemistry of organic and organometallic compounds. London: Academic Press 1970
[3] Connor, J. A., Skinner, H. A., Virmani, Y.: J.C.S. Faraday I *68*, 1754 (1972)
[4] Connor, J. A., Skinner, H. A., Virmani, Y.: Faraday Symposium Chem. Soc. *8*, 18 (1973)
[5] Adedeji, F. A., Brown, D. L. S., Connor, J. A., Leung, M. L., Paz-Andrade, M. I., Skinner, H. A.: J. Organomet. Chem. *97*, 221 (1975)
[6] Pittam, D. A., Pilcher, G., Barnes, D. S., Skinner, H. A., Todd, D.,: J. Less-Common Metals *42*, 217 (1975);
Shuman, M. S., Chernova, V. I., Zakharov, V. V., Rabinovich, I. B.: Trudy Khim, Khimtechnol. Gorky *1* (31), 78 (1974);
Cotton, F. A., Fischer, A. K., Wilkinson, G.: J. Amer. Chem. Soc. *78*, 5168 (1956)
[7] Telnoi, V. I., Rabinovich, I. B., Gribov, B. G., Pashinkin, A. S., Salamatin, B. A., Chernova, V. I.: Zh. Fiz. Khim. *46*, 802 (1972)
[8] Connor, J. A., Skinner, H. A., Virmani, Y.: J. C. S. Faraday I *69*, 1218 (1973)
[9] Brown, D. L. S., Connor, J. A., Skinner, H. A.: J. C. S. Faraday I *71*, 699 (1975)
[10] Brown, D. L. S., Connor, J. A., Leung, M. L., Paz-Andrade, M. I., Skinner, H. A.: J. Organometallic Chem. *110*. 79 (1976)
[11] Cotton, F. A., Fischer, A. K., Wilkinson, G.: J. Amer. Chem. Soc. *81*, 800 (1959)
[12] Good, W. D., Fairbrother, D. M., Waddington, G.: J. Phys. Chem. *62*, 853 (1958)
[13] Barnes, D. S., Pilcher, G., Pittam, D. A., Skinner, H. A., Todd, D., Virmani, Y.: J. Less-Common Metals *36*, 177 (1974)
[14] Barnes, D. S., Pilcher, G., Pittam, D. A., Skinner, H. A., Todd, D.: J. Less-Common Metals *38*, 53 (1974)
[15] Bidinosti, D. R., McIntyre, N. S.: Chem. Comm. *1966*, 555
[16] Brown, D. L. S., Connor, J. A., Skinner, H. A.: J. Organometallic Chem. *81*, 403 (1974)
[17] Junk, G. A., Svec, H. J.: J. Chem. Soc. (A) *1970*, 2102
[18] Chernova, V. I., Sheiman, M. S., Rabinovich, I. B., Syrkin, V. G.: Trudy Khim, Khimtechnol Gorky *1973*, 143
[19] Gardner, P. J., Cartner, A., Cunninghame, R. G., Robinson, B. H.: J. C. S. Dalton *1975*, 2582
[20] Fischer, A. K., Cotton, F. A., Wilkinson, G.: J. Amer. Chem. Soc. *79*, 2044 (1957)
[21] Abel, E. W., Stone, F. G. A.: Quart. Rev. *23*, 325 (1969)
[22] Forster, A., Johnson, B. F. G., Lewis, J., Matheson, T. W., Robinson, B. H., Jackson, W. G.: J. C. S. Chem. Comm. 1042 (1974);
Cotton, F. A., Hunter, D. L.: Inorg. Chim. Acta *11*, L9 (1974)
[23] Gurvich, L. V., Karachievtziev, G. V., Kondratiev, V. N., Lebedev, Yu. A., Medvedev, V. A., Potapov, V. K., Hodiev, Yu. S.: Dissociation energies of chemical bonds. Ionization potentials and electron affinities. Moscow: Nauka 1974
[24] Rutner, E., Haury, G. L.: J. Chem. Eng. Data *19*, 19 (1974) (nickel);
Cocke, D. L., Gingerich, K. A.: J. Chem. Phys. *57*, 3654 (1972);
Piacente, V., Balducci, G., Bardi, G.: J. Less-Common Metals *37*, 123 (1974) (rhodium);
Kordis, J., Gingerich, K. A., Seyse, R. J.: J. Chem. Phys. *61*, 5114 (1974) (gold)
[25] Donohue, J.: The structures of the elements. New York: Wiley 1974
[26] Oberteufer, J. A., Ibers, J. A.: Acta Cryst. *26B*, 1499 (1970)
[27] McIntosh, D., Ozin, G. A.: J. Amer. Chem. Soc. *98*, 3167 (1976)
[28] Kundig, E. P., McIntosh, D., Moskovits, M., Ozin, G. A.: J. Amer. Chem. Soc. *95*, 7234 (1973)

29) Johnson, B. F. G., J. C. S. Chem. Comm. *211*, 703 (1976)
30) Joyner, R. W., Roberts, M. W.: Chem. Phys. Letters *29*, 447 (1974)
31) Brennan, D., Hayes, F. H.: Phil. Trans. Roy. Soc. *258A*, 347 (1965);
Brennan, D., Hayward, D. O.: Phil. Trans. Roy. Soc. *258A*, 375 (1965)
32) Ertl, G.: Angew. Chem. *88*, 423 (1976); Internat. Edn. *15*, 391 (1976)
33) Madey, T. E., Menzel, D.: Japan J. Appl. Phys. Suppl. 2. Part 2. *1974* 229;
Fuggle, J. C., Madey, T. E., Steinkilberg, M., Menzel, D.: Surf. Sci. *52*, 521 (1975)
34) Ertl, G., Koch, J.: Z. Naturforschung *25A*, 1906 (1970)
35) Christmann, K., Ertl, G.: Z. Naturforschung *28A*, 1144 (1973)
36) Comrie, C. M., Weinberg, W. H.: J. Chem. Phys. *64*, 250 (1976)
37) Conrad, H., Ertl, G., Knözinger, H., Kuppers, J. Latta, E. E.: Chem. Phys. Letters. *42*, 115 (1976)
38) Thomas, J. L., Brown, K. T.: J. Organometallic Chem. *111*, 298 (1976)
39) Werner, R. P. M., Filbey, A. H., Manastryskyj, S. A.: Inorg. Chem. *3*, 298 (1964)
40) Werner, R. P. M., Podall, H. E.: Chem. Ind. (London) *1961*, 144;
Ellis, J. E., Davison, A.: Inorg. Syntheses *16*, 68 (1976)
41) Calabrese, J., Dahl, L. F., Chini, P., Longoni, G., Martinengo, S.: J. Amer. Chem. Soc. *96*, 2614 (1974);
Calabrese, J., Dahl, L. F., Cavalieri, A., Chini, P., Longoni, G., Martinengo, S.: J. Amer. Chem. Soc. *96*, 2616 (1974)
42) Battiston, G., Sbrignadello, G., Bor, G., Connor, J. A.: J. Organometallic Chem. *131*, 445 (1977)
43) Baerends, E. J., Ros, P.: J. Electron Spectroscopy *7*, 69 (1975);
Hillier, I. H., Saunders, V. R.: Mol. Phys. *22*, 1025 (1971)
44) Connor, J. A.: J. Organometallic Chem. *94*, 195 (1975)
45) Bassett, P. J., Higginson, B. R., Lloyd, D. R., Lynaugh, N., Roberts, P. J.: J. C. S. Dalton *1974*, 2316
46) Kiser, R. W., Krassoi, M. A., Clark, R. J.: J. Amer. Chem. Soc. *89*, 3653 (1967)
47) Brown, D. L. S., Connor, J. A., Skinner, H. A.: J. C. S. Faraday I *70*, 1649 (1974)
48) Bruce, M. I.: Adv. Organometallic Chem. *6*, 273 (1968)
49) Davidson, P. J., Lappert, M. F., Pearce, R.: Chem. Rev. *76*, 219 (1976);
Schrock, R. R., Parshall, G. W.: Chem. Rev. *76*, 243 (1976)
50) Adedeji, F. A., Connor, J. A., Skinner, H. A., Galyer, L., Wilkinson, G.: J. C. S. Chem. Comm. *1976*, 159
51) Lappert, M. F., Patil, D. S., Pedley, J. B.: J. C. S. Chem. Comm. *1975*, 830
52) Bradley, D. C., Hillyer, M. J.: Trans. Faraday Soc. *62*, 2367 (1966);
Shaulov, Y. K., Genchel, V. G., Aizatullova, R. M., Petrova, N. N.: Russian J. Phys. Chem. *46*, 1360 (1972);
Genchel, V. G., Volchkova, E. A., Aizatullova, R. M. Shaulov, Y. K.: Russian J. Phys. Chem. *47*, 643 (1973)
53) Telnoi, V. I., Rabinovich, I. B., Kozyrkin, B. I., Salamatin, B. A., Kiryanov, K. V.: Doklady A.N.S.S.S.R. Ser. Khim. *205*, 364 (1972)
54) Adedeji, F. A., Chisholm, M. H., Connor, J. A., Skinner, H. A.: unpublished work (1976)
55) Bradley, D. C. and Hillyer, M. J.: Trans. Faraday Soc. *62*, 2382 (1966)
56) Schrock, R. R., Meakin, P.: J. Amer. Chem. Soc. *96*, 5288 (1974)
57) Mertis, K., Wilkinson, G.: J. C. S. Dalton *1976*, 1488
58) Burgess, J., Fraser, C. J. W., Haigh, I., Peacock, R. D.: J. C. S. Dalton *1973*, 501
59) Porte, H. A., Greenberg, E., Hubbard, W. N.: J. Phys. Chem. *69*, 2308 (1965)
60) Anderson, R., Carmona Guzman, E., Mertis, K., Sigurdsson, E., Wilkinson, G.: J. Organometallic Chem. *99*, C19 (1975)
61) Pilcher, G., Ware, M. J., Pittam, D. A.: J. Less-Common Metals *42*, 233 (1975)
62) Schrock, R. R.: J. Amer. Chem. Soc. *96*, 6796 (1974)
63) Schrock, R. R.: J. Amer. Chem. Soc. *97*, 6577 (1975)
64) Connor, J. A., Riley, P. I.: J. C. S. Chem. Comm. *1976*, 149
65) Reger, D. L., Culbertson, E. C.: J. Amer. Chem. Soc. *98*, 2789 (1976)

66) Telnoi, V. I., Rabinovich, I. B.: Trudy Khim Khimtechnol. Gorky. 12 (1972); see also di Domenico, A., Harland, P. W., Franklin, J. L.: J. Chem. Phys. *56*, 5299 (1972); Furuyama, S., Golden, D. M., Benson, S. W.: Int. J. Chem. Kinetics *3*, 237 (1971)
67) Wilkinson, G., Pauson, P. L., Cotton, F. A.,: J. Amer. Chem. Soc. *76*, 1970 (1954); Roberts, J. S., Skinner, H. A.: Trans Faraday Soc. *45*, 339 (1949); Puttemans, J. P., Hanson, A.: Ind. Chim. (Brussels) *53*, 17 (1971)
68) Telnoi, V. I., Rabinovich, I. P., Latyaeva, V. N., Lineva, A. N.: Doklady A.N.S.S.S.R. Ser. Khim. *197*, 1348 (1971)
69) Telnoi, V. I., Kiryanov, K. V., Ermolaev, V. E., Rabinovich, I. B.: Doklady A.N.S.S.S.R. Ser. Khim. *220*, 1088 (1975); Trudy Khim. Khimtechnol. Gorky *4*, 3 (1975)
70) Telnoi, V. I., Rabinovich, I. B., Tikhonov, V. D., Latyaeva, V. N., Vyshinskaya, L. I., Razuvaev, G. A.: Doklady A.N.S.S.S.R. Ser Khim. *174*, 1374 (1967)
71) Guggenberger, L. J., Tebbe, F. N.: J. Amer. Chem. Soc. *98*, 4137 (1976)
72) Gard, E., Haaland, A., Novak, D. P., Seip, R.: J. Organometallic Chem. *88*, 181 (1975)
73) Atwood, J. L., Hunter, W. E., Alt, H., Rausch, M. D.: J. Amer. Chem. Soc. *98*, 2454 (1976)
74) Devyatykh, G. G., Rabinovich, I. B., Telnoi, V. I., Borisov, G. K., Zyuzina, L. F.: Doklady A.N.S.S.S.R., Ser. Khim. *217*, 609 (1974)
75) Atwood, J. L., Smith, K. D.: J. Amer. chem. Soc. *95*, 1488 (1973)
76) Fischer, A. K., Cotton, F. A., Wilkinson, G.: J. Phys. Chem. *63*, 154 (1959)
77) Fischer, E. O., Schreiner, S., Reckziegel, A.: Chem. Ber. *94*, 258 (1961)
78) Fischer, E. O., Reckziegel, A.: Chem. Ber. *94*, 2204 (1961)
79) Telnoi, V. I., Rabinovich, I. B., Umilin, V. A.: Doklady A.N.S.S.S.R. Ser. Khim. *209*, 127 (1973)
80) Herberich, G. E., Müller, J.: J. Organometallic Chem. *16*, 111 (1969)
81) Cais, M., Lupin, M. S.: Adv. Organometallic Chem. *8*, 211 (1970)
82) Müller, J. and D'Or, L.: J. Organometallic Chem. *10*, 313 (1967); see also Gaivoronski, P. E., and Larin N. V.: Trudy Khim. Khimtechnol. Gorky *3*, 74 (1973)
83) Dahl, L. F., Martell, C., Wampler, D. L.: J. Amer. Chem. Soc. *83*, 1761 (1961)
84) Brown, D. L. S. Cavell, S., Connor, J. A., Leung, M. L., Skinner, H. A., unpublished work (1973–1975)
85) Pribula, A. J., Drago, R. S.: J. Amer. Chem. Soc. *98*, 2784 (1976)
86) Wojcicki, A.: Adv. Organometallic Chem. *11*, 87 (1973)
87) Calderazzo, F., Cotton, F. A.: Inorg. Chem. *1*, 30 (1962)
88) Blake, D. M., de Faller, J., Chung, Y. L., Winkelman, A.: J. Amer. Chem. Soc. *96*, 5568 (1974); Blake, D. M., Winkelman, A., Chung, Y. L.: Inorg. Chem. *14*, 1326 (1975)
89) Meester, M. A. M., Vriends, R. C. J., Stufkens, D. J., Vrieze, K.: Inorg. Chim. Acta. *19*, 95 (1976)
90) Bleijerveld, R. H. T., Vrieze, K.: Inorg. Chim. Acta. *19*, 195 (1976)
91) Aue, D. H., Webb, H. M., Bowers, M. T.: J. Amer. Chem. Soc. *98*, 318 (1976)
92) Adedeji, F. A., Connor, J. A., Demain, C. P., Martinho Simoes, J. A., Skinner, H. A.: unpublished work (1976)
93) Corderman, R. R., Beauchamp, J. L.: J. Amer. Chem. Soc. *98*, 3999 (1976)
94) Tolman, C. A., Reutter, D. W., Seidel, W. C.: J. Organometallic Chem. *117*, C30 (1976)
95) Sneeden, R. P. A.: Organochromium compounds. New York: Academic Press 1975
96) Müller, J.: J. Organometallic Chem. *18*, 321 (1969)
97) Huttner, G., Krieg, B.: Chem. Ber. *105*, 3437 (1972)
98) Scotti, M., Werner, H.: Helv. Chim. Acta *57*, 1234 (1974)
99) Khand, I. U., Knox, G. R., Pauson, P. L., Watts, W. E.: Chem. Comm. 36 (1971). J.C.S. Perkin I, *1973*, 975
100) Evstigneeva, E. V., Shmyreva, G. O.: Russian J. Phys. Chem. *39*, 529 (1965)
101) Fischer, E. O., Werner, H.: Metall π- complexes. Vol. 1. Complexes with di- and oligo-olefinic ligands. Amsterdam: Elsevier 1966; Herberhold, M.: Metall π- complexes. Vol. 2. Complexes with mono- olefinic ligands. Amsterdam: Elsevier Part I. 1972, Part II. 1974

102) Pappenheimer, W.: J. Amer. Chem. Soc. *98,* 2779 (1976) and references
103) Tolman, C. A.: J. Amer. Chem. Soc. *96,* 2780 (1974)
104) McNaughton, J. L., Mortimer, C. T., Burgess, J., Hacker, M. J., Kemitt, R. D. W.: J. Organometallic Chem. *71,* 287 (1974)
105) Cramer, R.: J. Amer. Chem. Soc. *94,* 5681 (1972)
106) Saalfeld, F. E., McDowell, M. V., DeCorpo, J. J., Berry, A. D., MacDiarmid, A. G.: Inorg. Chem. *12,* 48 (1973)
107) Saalfeld, F. E., McDowell, M. V., Gondal, S. K., MacDiarmid, A. G.: J. Amer. Chem. Soc. *90,* 3684 (1968)
108) Bronshtein, Yu. E., Gankin, V. Y., Prinkin, D. P., Rudovskii, D. M.: Russian J. Phys. Chem. *40,* 802 (1966)
109) Foster, J. S., Beauchamp, J. L.: J. Amer. Chem. Soc. *97,* 4808 (1975)
110) Foster, J. S., Beauchamp, J. L.: J. Amer. Chem. Soc. *97,* 4814 (1975)
111) Corderman, R. R., Beauchamp, J. L.: Inorg. Chem. *15,* 665 (1976)
112) Eley, D. D.: Disc. Faraday Soc. *8,* 34 (1950)
113) Tolman, C. A.: Chem. Rev. *77* (1977)
114) Reed, P. D., Comrie, C. M., Lambert, R. M.: Surf. Sci. *59,* 33 (1976)
115) Alemdaroglu, N. H., Penninger, J. M. L., Oltay, E.: Monatshefte *107,* 1043 (1976)
116) Fawcett, J. P., Poë. A. J., Sharma, K. S.: J. Amer. Chem. Soc. *98,* 1401 (1976)
117) Bidinosti, D. R., McIntyre, N. S.: Canad. J. Chem. *48,* 593 (1970)
118) Fawcett, J. P., Poë. A. J.: J. C. S. Dalton *1976,* 2039
119) Haines, L. I. B., Poë. A. J.: J. Chem. Soc. A. *1969,* 2826
120) Krause, J. R., Bidinosti, D. R.: Canad. J. Chem. *53,* 628 (1975)
121) Cutler, A. R., Rosenblum, M.: J. Organometallic Chem. *120,* 87 (1976)
122) Barrett, P. F., Poë. A. J.: J. Chem. Soc. A. *1968,* 429

Received October, 11, 1976

The Vibrational Spectra of Metal Carbonyls

Prof. Sidney F. A. Kettle*

Department of Chemistry, Northwestern University, Evanston, Illinois 60201, U.S.A.

Table of Contents

* On leave from the School of Chemical Sciences, The University of East Anglia, Norwich, NR4 7TJ U.K.

1. Introduction

The most recent fairly comprehensive review of the vibrational spectra of transition metal carbonyls is contained in the book by Braterman[1]. This provides a literature coverage up to the end of 1971 and so the subject of the present article is the literature from 1972 through to the end of 1975. Inevitably, some considerable selectivity has been necessary. For instance, a considerable number of largely preparative papers are not included in the present article. Tables A-E provide a general view of the work reported in the period. Table A covers spectral reports and papers for which topics related purely to vibrational analysis are not the main objective. Papers with the latter more in view are covered in Table C. Evidently, the division between the two is somewhat arbitrary. Other tables are devoted to papers primarily concerned with the spectra of crystalline samples – Table B – to reports of infrared and Raman band intensities – Table D and sundry experimental techniques or observations – Table E. Papers on matrix isolated species, which are covered elsewhere in this volume, are excluded.

Several aspects are selected for particular review in the following pages. Inevitably, those chosen tend to reflect the reviewers own interests and prejudices but are also topics for which no review has previously been carried out.

2. General Developments

The period under review has seen a small, but apparently real, decrease in the annual number of publications in the field of the vibrational spectroscopy of transition metal carbonyls. Perhaps more important, and not unrelated, has been the change in perspective of the subject over the last few years. Although it continues to be widely used, the emphasis has moved from the simple method of ν(CO) vibrational analysis first proposed by Cotton and Kraihanzel[2], which itself is derived from an earlier model[4], to more accurate analyses. One of the attractions of the Cotton-Kraihanzel model is its economy of parameters, making it appropriate if under-determination is to be avoided. Two developments have changed this situation. Firstly, the widespread availability of Raman facilities has made observable frequencies which previously were either only indirectly or uncertainly available. Not unfrequently, however, these additional Raman data have been obtained from studies on crystalline samples, a procedure which, in view of the additional spectral features which can occur with crystalline solids (*vide infra*), must be regarded as questionable. The second source of new information has been studies on isotopically-labelled species. ^{13}C is present in 1.1% natural abundance and so approximately n% of a $M(CO)_n$ complex is isotopically labelled. General infrared instrumental improvements have made studies on such natural-abundance species perfectly feasable, for most ν(CO) infrared peaks are rather narrow in hydrocarbon solvents. The commercial availability of ^{13}CO and C^{18}O (substitution by either of which has very similar frequency perturbation consequences) has led to studies of isotopically enriched species. Perhaps because of the emphasis on studies of ν(CO) vibrations, little work has been reported involving metal isotope studies[A57]. Isotopic studies are frequently confined to infrared spectra. Firstly, the Raman spectroscopy of solutions usually demands more sample than does infrared. Secondly, Raman studies are frequently carried out in a chlorinated solvent-despite the dangers of reaction[E1] – because these afford a higher solubility than the otherwise preferred hydrocarbons. In non-hydrocarbon solvents peaks tend to be somewhat broadened and isotopic features may be less well resolved. A further problem with such solvents is the frequency shifts which can occur compared to those obtained in hydrocarbons, thereby perhaps making questionable the use of Raman data from such solvents in vibrational analyses.

Rather than attempt a complete review of all the work in this field the present article will summarize the developments in the vibrational analysis of one particular species, $Mn(CO)_5Br$. The infrared spectrum of this molecule was first reported by Wilson[5] and this and the Raman spectra have been subject to many investigations since. It therefore provides a good bird's-eye view of the development of the subject.

Surprisingly, the earliest study made of a vibrational spectrum of $Mn(CO)_5Br$ appears to have been Cable and Sheline's attempt to record its Raman spectrum[6]. Unfortunately, the use of a mercury arc led to photodecomposition, a problem which was also encountered by early workers using a laser source[7]. This report mentions a strong band at 2079.5 cm^{-1}, a band also found by Davidson and Faller[8]. They found a total of four bands in the ν(CO) region in the Raman spectrum of the solid. Similarly, although the earliest reports of the ν(CO) infrared spectrum mention two strong bands in chloroform solution[9], improved resolution (and greater instrumental frequency accuracy) showed the presence of at least four bands in solution spectra, of which one originates in a ^{13}CO-containing species[10]. These authors used a simple model of coupled ν(CO) oscillating dipoles to interpret both their frequency and intensity data. Their assignment of the highest frequency mode as an in-phase stretching of equatorial carbonyl groups, the next as their degenerate out-of-phase vibration and the lowest as an axial ν(CO) stretch is identical to that deduced by Orgel[11]. Orgel recognized the correlation between the modes of an octahedral hexacarbonyl and those of C_{4v} species such as $Mn(CO)_5Br$.

No doubt it was the absence of detailed infrared data which prevented both Cotton and Kraihanzel[2,3] and Poilblanc and Bigorgne[12], who studied $C_{4v}M(CO)_5L$ systems, from including $Mn(CO)_5Br$ in their work. This compound was subsequently studied by Cotton, Musco and Yagupsky[39]; although, following a demonstration of its near applicability to $CF_3Mn(CO)_5$, they reported an analysis based on the postulate that $k_{ee}^c = k_{ea}$ (Fig. 1). Although this approximation was not really necessary, because they were not underdetermined (they had available Raman, ^{13}C isotopic

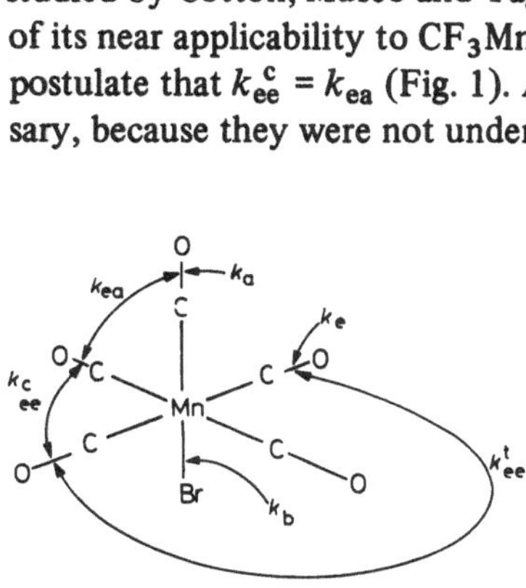

Fig. 1. Force and interaction constants in a $M(CO)_5X$ system (see Table 1)

and overtone/combination data), they concluded that this postulated relative magnitude of interaction constants provided a reasonable first approximation. It is therefore of interest to see whether this view persists in a more extensive analysis[13].

A somewhat more detailed analysis of $Mn(CO)_5Br$ is, for instance, that of Kaesz, Bau, Hendrickson and Smith[14], who confined their analyses to the ν(CO) region but by including data on isotopic species (and also carrying out ^{13}CO enrichment studies) obtained the data given in column 2 of Table 1. It can be seen that they conclude that $k^t_{ee} > k_{ea} > k^c_{ee}$ and that the ratios of k_{trans}/k_{cis} are 2:3 and 1:4, in reasonable agreement with the Cotton-Kraihanzel model[2, 3].

A much more detailed analysis has appeared in the period under review[15]. Ottesen, Gray, Jones and Goldblatt included metal-carbon stretching and deformation modes in their analysis; they prepared and reported data on $Mn(^{12}CO)_5Br$, $Mn(^{13}CO)_5Br$ and $Mn(C^{18}O)_5Br$. A selection of their results is detailed in Table 1.

Table 1. Force and interaction constants for $Mn(CO)_5Br$

m. dynes $Å^{-1}$	Ref.[39]	Ref.[14]	Ref.[78]	1)	Ref.[16]	Ref.[15]
k_a	16.22	16.35	16.392	16.328	–	16.792
k_e	17.44	17.41	17.458	17.453	–	17.884
k_{ea}	0.21	0.305	0.310	0.262	–	0.204
k^c_{ee}	0.21	0.186	0.178	0.192	–	0.094
k^t_{ee}	0.45	0.432	0.429	0.437	–	0.073
k_b	–	–	–	–	0.91 to 1.58	1.318

1) G. Bor unpublished results (cos β method).

In particular, it is to be noted that $k_{ea} > k^c_{ee} > k^t_{ee}$ so that the ratios k_{trans}/k_{cis} are 0 · 8 and 0 · 4, well away from the Cotton-Kraihanzel values. These workers were also able *inter alia* to improve on an earlier estimate of the Mn-Br stretching force constant[16]. Although this latter paper by Clarke and Crosse is one of the few to discuss lowfrequency apectra, they also report the ν(CO) Raman spectrum. Like earlier workers[8] they used a crystalline sample and found five features, four of which are strong. In contrast, the solution data of Ottesen, Gray, Jones and Goldblatt show only three non-isotopic features. This illustrates the difficulty- and danger- of attempting to incorporate Raman data from crystalline samples in a vibrational analysis of an isolated molecule. In general, the error is likely to be greater the greater the concentration (including local high concentrations) of CO groups within a crystal. This problem had previously been explicitly recognized by Cotton, Musco and Yagupsky[39]. The comparison in Table 1, whilst raising no question against the use of group frequency [here ν(CO)] vibrational analyses as an aid to assignment does appear to raise doubts about their use as measures of absolute bonding strengths[17]. Not that even detailed analyses are without their problems. How, without additional data, is one to determine whether, in the limit, the deformation force constant in an AB_2 system tells us something about the relative slopes of components in a Walsh diagram or about repulsive forces between the two B atoms?

There is a variety of problems associated with the details of the ν(CO) vibrational spectra of $Mn(CO)_5Br$. For instance, the formally infrared inactive B_2 mode is almost invariably seen as a weak feature. For most C_{4v} pentacarbonyls this is a result of asymmetry in the sixth ligand attached to the metal[18] but in the case of $Mn(CO)_5Br$ and similar species there seems to be no alternative to the involvement of one quantum of a low frequency degenerate mode leading to a non-totally symmetric vibrational ground state and consequent modification of the selection rules[19]. More commonly, however, the situation is the opposite. The doubly degenerate ν(CO) mode of E symmetry is formally Raman active but appears only weakly. Similarly, the feature associated with the low energy, totally symmetric, ν(CO), peak is not strongly polarized. Such observations have led Edgell to propose that there exists within $Mn(CO)_5Br$ a hidden symmetry. In its original presentation this theory was particularly appropriate to molecules with fragments displaying characteristic group frequencies[20], the potential and kinetic energy matrices being diagonalized by a set of modes which includes some which may be localized in molecular fragments. It may therefore be appropriate to describe a local symmetry (L) (*e.g.* C_{3v} for a methyl group). Further, there will exist a group, I, of permutational operators which formally interchange atoms between such localized groups. There is no formal requirement that all elements of I connect chemically, as opposed to vibrationally, similar atomic groups. Edgell then demonstrates that is is appropriate to consider the extra-molecular point group symmetry of the direct product group $G = L \times I$ (although the factor group G/I may be adequate; it is isomorphic to L). In the case of $Mn(CO)_5Br$, Edgell considers, rather, the permutational (symmetric) group P_5, the molecular point group C_{4v} and, instead of the usual symmetric group notation[21], uses one designed to show the relationship between the two groups (which is one of correlation rather than factor). Edgell concludes that it is this, correlated, group which provides an explanation of the observed vibrational features. It is clear from Table 1 that the five CO groups in $Mn(CO)_5Br$ are not vibrationally equivalent and so the S_5 group, itself, not appropriate. Equally, it is clear that the spectral explanations offered by Edgell are quite different from the generally accepted alternative, the correlation with O_h argument of Orgel[11]. For instance, the argument that the $\nu(CO)E$ modes are weak in the Raman because of their correlation with the T_{1u} modes in O_h would persist if the axial CO group were removed to give a planar $Mn(CO)_4$ unit whereas Edgell's argument would have to be recast. There is one hitherto unexplained phenomenon for which Edgell offers an explanation; the fact that the Raman peak associated with the, low-energy, axial CO stretching vibration is little polarized. Here, though, an alternative explanation is not hard to find. A vibration is totally polarized when it has spherical symmetry (or at least cubic). Then the moment induced by the electric vector of the incident light is parallel to that vector resulting in complete polarization of the scattered radiation. The A_{1g} ν(CO) mode of the hexacarbonyls provides a pertinent example[D8]. Suppose we have a set of coupled vibrators, equidistant from some origin. Then it must be possible to express the basis functions for the vibrations in terms of spherical harmonics, for the former are orthogonal and the latter comprise a complete set. The polarization of a totally symmetric vibration will be determined by its overlap with the spherically symmetrical term which may be taken as $r^2 = x^2 + y^2 + z^2$. Because of the orthogonality of the vibrational functions, if one totally symmetric vibration is highly polarized (*i.e.* has a large overlap) a second will not be. In the present case the high energy ν(CO)A_1 mode transforms almost precisely as $x^2 + y^2$ (it is weak in the IR) and is highly polarized. The low energy A_1 mode transforms, rather, as z and so is relatively strong in the IR and displays little polarization in the Raman. This argument is closer to that of Edgell than that concerning E-mode activity discussed above[79)].

A detailed analysis of the factors influencing the relative infrared intensities of the two (CO)A_1 modes has been given by Manning and Miller[40)]. Again, this analysis is quite different to that of Edgell. It is difficult to envisage a definitive test of Edgell's approach. S_3 and S_4 are isomorphic to C_{3v} and T_d. Perhaps application of S_4 to a tetracarbonyl with effective C_{4v} symmetry, S_6 to an octahedral hexacarbonyl or S_5 to a trigonal biprismatic pentacarbonyl would provide further insight.

In the case which we have just considered the permutations involved were algebraic rather than physical. Rather different is the work of Bor[22)]. He showed that the infrared spectra of $Co(CO)_4 \cdot M(CO)_5$, M = Mn, Re[23)] and M = Tc[24)] can only be interpreted in terms of a model which allows free rotation about the metal-metal bond. This is perhaps not too surprising since the juxtaposition of the C_{4v} and C_{3v} units requires a twelve-fold barrier to free rotation (assuming no distortion of the units) and such a barrier is likely to be low. It is recognized that the vibrational point group of an equilibrium configuration may be used to describe the vibrational features of what is, intrinsically, a molecule distorted away from its equilibrium configuration, by nature of the correspondence between cosets of the complete permutation-inversion group and the molecular point group[25, 26)]. A formal application of such an approach to $Co(CO)_4 \cdot M(CO)_5$ species would probably provide insight into the detailed origin of the spectral simplifications explained by Bor.

3. Raman Intensities

There have been extensive studies of ν(CO) infrared intensities (but fewer extending to other spectral regions, notwithstanding the fact that mixing with the ν(MC) modes should be included in analyses) in some part motivated by the recognition that such measurements enable often surprisingly accurate predictions of the angle between adjacent CO groups (which, preferably, are symmetry-related)[D5]. In contrast, there are very few measurements of Raman intensities and no two groups agree on the preferred method of analysis.

Experimentally, several precautions must be taken if reliable Raman data are to be obtained from solution studies. Firstly, the instrumental slit-width should be appreciably smaller than the half-width of the band to be studied. This means that slits wider than 2 cm^{-1} are to be avoided. Secondly, photolytic decomposition of the sample and local boiling of the solvent have also to be avoided. Careful choice of laser frequencies, use of a low incident power and, if necessary, sample spinning are indicated. The need for a relatively high solute concentration usually means that there is little choice of solvent. Particularly for coloured samples the presence of a vestigal resonance Raman effect must be tested by measurements with a variety of laser frequencies. Because Raman spectroscopy is a single beam technique the experimental data have to be corrected for the wavelength dependent efficiencies of the optics and the photomultiplier. Simplest is to calibrate the spectrometer with a sub-standard white light source. Typically this standard would be a calibrated car headlamp-type of bulb, held in the attitude and holder in which it was calibrated and run at its calibration current. A variety of techniques may be used to measure spectra: digitization, cut-and-weight, planimetry. The last named has the advantage of simplicity and speed: high accuracy is not necessary since the scatter on measurements on different spectra may be 5%. The rate-meter-type of plotting usually employed makes data processing simple; wing corrections are not usually employed. Absolute Raman intensities may, in principle, be made using an internal calibrant.

The earliest measurements of the Raman spectra of metal carbonyls, using arc[27)] or laser[28)] excitation reported only frequency data. The first report of intensities and their interpretation is to be found in the work of Terzis and Spiro[29)]. These authors studied the ν(CO) and $\nu(MC) A_{1g}$ peaks of $Cr(CO)_6$, $Mo(CO)_6$ and $W(CO)_6$, interpreting their results with a δ-function model. In such models a molecule is represented as a smoothly varying potential in space except at nuclear positions, where there is a spike. The model is therefore akin to an electron in a box. By suitable parametrization, the model can be made to give good predictions but its applicability to phenomena outside its parametrization is uncertain. This model was first introduced by Long and Plane[30)]. In it there is no formal contribution from non-bonding electrons and the derived polarizability tensor is assumed to consist of along-bond polarizability changes only. The model involves valence state electronegativities in a somewhat arbitrary way and, through an electronegativity difference between the atoms in a bond, the covalent bond character. For the C-O bond a bond order of $4 \cdot 2$ to $4 \cdot 4$ is indicated by Terzis and Spiro's results. This bond order is rather difficult to believe and they attributed it to a breakdown of the bond polarizability theory. An alternative explanation is a breakdown of the δ-model itself. Chantry and Plane[31)] had previously doubted the applicability of a bond polarizability model to explain Raman intensities of totally symmetric modes in some cyanide complexes in comparison with others in which it seemed appropriate. Their argument, however, seems to be based on an expectancy of constancy of derived intensities. They noted that 'anomalous' behaviour was associated with an anomalous ultraviolet spectrum, a feature which may not be surprising in view of the importance of electric-dipole allowed electronic transitions in determining Raman intensities[32)].

In 1972 two further approaches to the Raman intensities in metal carbonyl spectra were published[D6, D8, D9], one of which had been the subject of two preliminary communications[33, 34)]. Koenig and Bigorgne reported the absolute Raman intensities of the totally symmetric modes of the species $Ni(CO)_4$, $Co(CO)_4^-$, $Ni(CO)_{4-n}L_n$, L = PMe_3 (n = 1,2) and L = $(OMe)_3$ (n = 1, 2, 3). In the present review we shall be particularly concerned with their results for the Ni-C and C-O bonds but they also reported data for P-C and $\widehat{PCC}$. Like Terzis and Spiro, their data are presented in terms of a Wolkenstein, bond additivity, model[35–37)]. Rather than an analysis by a δ-potential model, Koenig and Bigorgne attempt to decompose the observed derived bond polarizabilities. They assume that components arise from the CO-bond stretch (dependent on bond order), on unpaired electron density on carbon and oxygen and on metal-carbon σ and π bonding. In setting up equations to determine these quantities, Koenig and Bigorgne admit to an element of empiricism. Thus, the C-O bond is assumed triple in CO and $Ni(CO)_4$ but double in $Ni(CO)[P(OMe)_3]_3$ and acetone. Similarly, the contribution from a Ni-C σ bond is assumed identical to that of a C-C. The results of this analysis lead to higher values for CO bond derived polarizabilities (by a factor of almost three) than are obtained from the raw data. The available absolute intensity data are summarized in Table 2.

Kettle, Paul and Stamper[33, 34) D8] are the only group who have reported intensity data for other than totally symmetric modes. They confined their attention to the ν(CO) region and looked at octahedral $M(CO)_6$, trigonal $RM(CO)_3$ and tetragonal $R'M(CO)_5$ species. The measurement of the intensities of several modes may enable

Table 2. Absolute raman intensities

Compound	Bond	$\bar{\alpha}'$	Model	Ref.
$Cr(CO)_6$	MC	1.05	δ function/	29)
	CO	1.74	Wolkenstein	29)
$Mo(CO)_6$	MC	1.21	δ function/	29)
	CO	1.44	Wolkenstein	29)
$W(CO)_6$	MC	1.54	δ function/	29)
	CO	0.92	Wolkenstein	29)
$Ni(CO)_4$	MC	0.88	Wolkenstein	D6
	CO	1.04		D6
$Ni(CO)_3PMe_3$	MC	0.67	Wolkenstein	D6
	CO	0.73		D6
$Ni(CO)_2(PMe_2)_2$	MC	~0.77	Wolkenstein	D6
	CO	0.36		D6
$Ni(CO)_3P(OMe)_3$	MC	~0.99	Wolkenstein	D6
	CO	0.98		D6
$Ni(CO)_2[P(OMe)_3]_2$	MC	0.68	Wolkenstein	D6
	CO	0.54		D6
$Ni(CO)[P(OMe_3)_3]_3$	MC	~0.42	Wolkenstein	D6
	CO	0.11		D6
$Co(CO)_4^-$	MC	0.80	Wolkenstein	D6
	CO	0.22		D6
$Ni(CO)_4$	CO	2.70	Wolkenstein (decomposed)	D6
$Ni(CO)_3PMe_3$	CO	2.19	Wolkenstein (decomposed)	D6
$Ni(CO)_2(PMe_3)_2$	CO	2.10	Wolkenstein (decomposed)	D6
$Ni(CO)_3P(OMe)_3$	CO	3.07	Wolkenstein (decomposed)	D6
$Ni(CO)_2(P(OMe)_3)_2$	CO	2.25	Wolkenstein (decomposed)	D6
$Ni(CO)(P(OMe)_3)_3$	CO	1.35	Wolkenstein (decomposed)	D6

not only the estimation of an average CO derived polarizability but also of the individual tensor components. In fact, Kettle *et al.* do not report absolute intensity data but only relative. Their data may, however, be combined with that in Table 2 to give absolute values using the relationship

$$\alpha' = {}^1/3\,(\alpha_1' + \alpha_2' + \alpha_3')$$

where the prime denotes a derivative with respect to r_{CO} and α_1, α_2, and α_3 are the three principal tensor derivatives. In their original report, Kettle, Paul and Stamper ignore the effects of $\nu(CO) - \nu(MC)$ mixing. More recently, these effects have been incorporated and found to produce only relatively small changes in the values originally reported, although providing the unexpected result that the ν(M-C) longitudinal and transverse bond derived polarizability components are also of opposite sign[38]. Using the original values one calculates for $Cr(CO)_6$, for instance, that $\alpha_{11} = 9 \cdot 7$ Å amu$^{-1/2}$ and $\alpha_\perp' = -2.2$ Å amu$^{-1/2}$. Although Kettle, Paul and Stamper made no attempt to interpret their data in terms of chemical bonding, it is tempting to relate the different signs for their polarizability derivatives to different patterns of σ and π bonding. Unfortunately, like the other workers in the field, they found no indication of a general transferability of their quantities. This, of course, is related to a somewhat variable Raman intensity pattern. Thus, whilst

$\frac{\alpha_\perp'}{\alpha_{11}'}$ was found to be approximately -0.23 for $Cr(CO)_6$, $Mo(CO)_6$ and $W(CO)_6$ values of -0.28 and -0.08 were found for $V(CO)_6^-$ and $Re(CO)_6^+$, respectively; spectra for the latter two compounds were obtained from the literature and, lacking the corrections applied to their own data, the resulting tensor ratios may well be less accurate. Even so, the factor of 3 which relates the ratio for the rhenium compound to the others demonstrates the lack of transferability. Whilst it could be argued that the data on metal carbonyls, taken together, could indicate, as has been suggested, the inapplicability of a bond polarizability model, it would seem to the reviewer that this is a premature conclusion. Firstly, it has yet to be established that there is a reason to expect bond polarizability derivatives to be transferable. Rather, insofar as they reflect the different patterns of electronic states in the various molecules a nonuniformity is to be expected. Only if the electronically excited states of each CO ligand are totally insulated from those of the rest of the molecule should real transferability exist. Secondly, a bond polarizability theory only postulates a 1 : 1 correspondence between a set of chemical bonds and a set of derived polarizability tensors. Each tensor may well contain components from bonds other than that with which it is isomorphic; the consequence of this will probably be that the bond tensor axes do not coincide with the chemically natural axes – for instance, there may not be a tensor axis precisely coincident with the bond axis. The bond tensor model provides a recipe for the generation of molecular from bond tensors by specifying appropriate rotation transformations. It is this aspect of the model which breaks down if bond tensor axes depart markedly from the chemically sensible ones. However, for most of the species which have been the subject of investigation, the orientation of tensor axes is symmetry, and not chemically, determined and for these the problem of mis-orientation does not arise.

$Mn(CO)_5Br$ bears a close structural relationship to the hexacarbonyls so that the values of the ratios of its ν(CO)-derived polarizability tensors are of interest. For the axial carbonyl the ratio $\alpha'_\perp/\alpha'_{11}$ is either −0.99 or −0.20 (the two roots of a quadratic equation obtained with neglect of L-matrix effects) compared with a value of ca. −0.23 for the hexacarbonyls[D8]. The derived polarizability tensor for each equatorial CO group is not cylindrically symmetrical; the two values of $\frac{\alpha'_\perp}{\alpha'_{11}}$ are either −0.27 and 0.36 or −3.76 and −1.81. Although a full transferability of ratios does not exist, the common occurence of a $\alpha'_\perp/\alpha'_{11}$ ratio of -0.23 ± 0.04 is suggestive that, whilst absolute magnitudes may have little transferability, ratios of derived polarizability elements may be more transferable. Even if this suggestion is true for similar species it seems that it can scarcely be generally true; thus for a range of arylchromium tricarbonyl species, for which the CO bands do not have cylindrical symmetry, the values of $\alpha'_\perp/\alpha'_{11}$ found were 0.11 and −0.78. Although transferable within this class of compounds these ratios are scarcely transferable with those discussed above.

4. Studies of Crystalline Materials

In studies of solutions of metal carbonyls the unit which is used as a basis for the explanation of vibrational spectra is the molecule. In studies on crystalline materials the corresponding building block is the unit cell. The reason for this is that the wavelength of radiation used (be it infrared or laser Raman) is large compared to inter-atomic distances. This means that the effect of the electric vector of the incident radiation will be essentially identical for all molecules in proximate unit cells which are related by a pure translation operation. What is excited is an in-phase vibration of all translationally-related molecules within the unit cells. This argument does not depend at all on the existence of vibrational coupling between adjacent molecules in the crystal. If such couplings are negligible then the spectral features observed in internal molecular mode spectral regions will correspond closely with those of the isolated molecule. The only complications arise from the possibility of multiple site occupancy within the crystal (*i.e.* symmetry unrelated molecules which, because of the consequent differences in local potential, will usually mean that there will be one peak associated with each site for a non-degenerate mode). If vibrational degeneracy occurs in the isolated molecule the crystallographic site symmetry will usually be appreciably lower than the molecular and a splitting of the degeneracy will occur[41)]. This is the so-called site-symmetry model.

All of these features persist when inter-molecular vibrational coupling cannot be neglected. Effectively, the unit cell is treated as a giant molecule and from symmetry-related vibrators in the molecules in the unit cell symmetry-adapted bases are projected out. Thus, in a unit cell containing just four vibrators the combinations would be (++++), (++−−), (+−+−) and +−−+). The different combinations give rise to different frequencies and spectral activities (so that coupling gives rise to a richer spectrum). This is the correlation field, the factor group or unit cell group approach (although care has to be taken with the last name since in using it it is sometimes, incorrectly, implied that a unique unit cell exists for a particular structure). In practice, rather than carry out a full analysis, although tables are available to assist in this[42)], it is simplest to use a correlation method[43)]. Many articles on factor group analyses are available[44–49)]; one which deals in a readable manner with the difficult problem of non-symmorphic space groups is to be recommended[50)]. More detailed analyses, which provide some insights into the way that the subject may develop, are available[51–57)]. Attempts are being made to carry out normal coordinate analyses of crystals, although the problem is much more difficult than for isolated molecules[58, 59)], and a computer program is available[60)].

There is no doubt that, in general, it is necessary to use a factor group approach to explain the internal mode vibrational spectra of transition metal carbonyls. Cases in which it is unnecessary are, indeed, rather rare. One example is provided by $(Ph_2C_2)_3W(CO)$, which shows two sharp ν(CO) features of identical intensity in both infrared and Raman, at coincident frequencies[61)]. This coincidence is strongly indicative of the absence of factor group coupling, and it seems reasonable to suppose that the two peaks correspond to the two crystallographically non-equivalent sites which are occupied in the crystal. This hypothesis is confirmed by ^{13}CO enrichment studies which show that two additional lower frequency, equal intensity and similarly separated peaks, grow in intensity with increasing ^{13}CO substitution, the infrared and Raman peaks again being coincident (Fig. 2). In contrast, in all other carbonyl

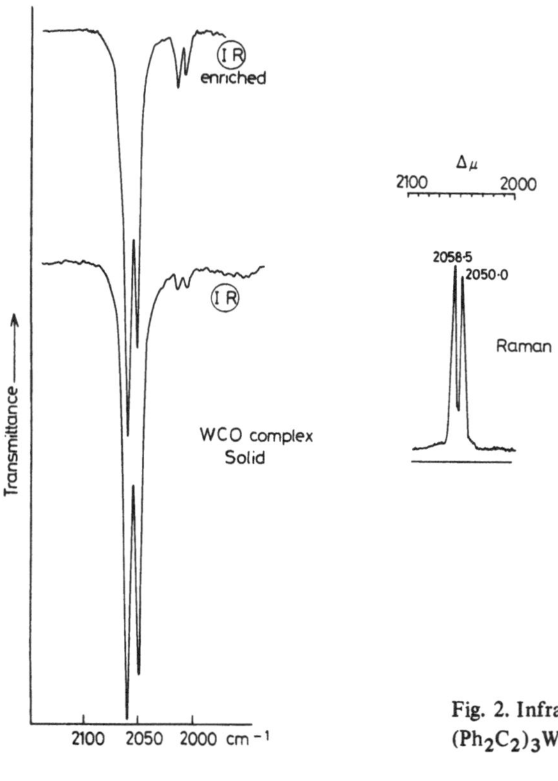

Fig. 2. Infrared and Raman spectra of crystalline $(Ph_2C_2)_3W(CO)$ in the 2000 cm^{-1} region

species which have been studied (and which have a much higher concentration of CO groups), it seems probable that factor group splitting has been observed in the ν(CO) region. Two comments are appropriate. Firstly, there has been no definite report of such splitting in either the ν(M-C) or δ (CMC) region, although factor group effects undoubtedly occur in the low frequency, lattice mode, region. Secondly, a study of either the infrared or Raman spectrum on its own may not be adequate to reach a conclusion on the presence or absence of factor group splitting. This is because it quite often happens that the site model provides an apparently acceptable explanation. Benzenetricarbonylchromium is a good example. Both infrared and Raman spectra in the ν(CO) region consist of three strong peaks. For either on its own it appears that the high energy peak could well be the molecular A_1 mode and the lower energy pair the degenerate molecular E mode, split by site symmetry effects. It is only when the infrared and Raman spectra are compared and it is found that no coincidences exist that it becomes apparent that there are six ν(CO) vibrational features (Fig. 3), as appropriate for a monoclinic unit cell containing six CO groups (Fig. 4)[62, 63)]. It is not always as simple as this. In particular, ν(CO) infrared spectra of crystalline samples can show very broad bands with an indeterminate number of components. The hexacarbonyls of chromium, molybdenum and tungsten provide examples of this but, fortunately, it is clear from the Raman spectrum that factor group effects occur (Fig. 5)[B5]. It is often the case than ν(CO) Raman spectra

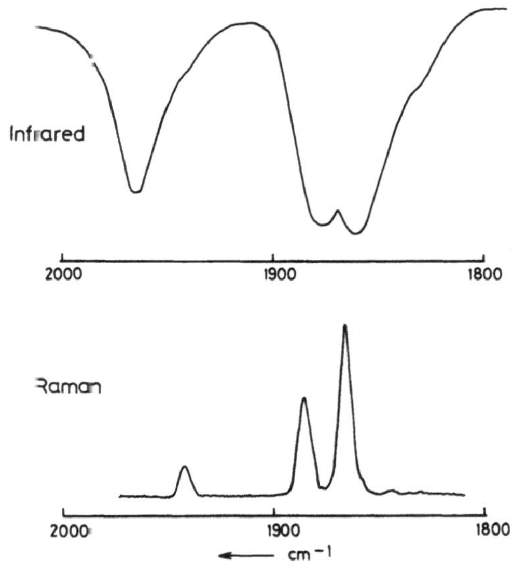

Fig. 3. Infrared and Raman spectra of crystalline η^6-$C_6H_6Cr(CO)_3$ in the 2000 cm^{-1} region

of crystalline samples are more informative than are infrared. Not only do the latter tend to be broad but also, when studied as KBr(or similar) discs, the spectrum obtained tends to be dependent upon the grinding given in making the disc. Increased grinding usually leads to band sharpening, which is evidently good, but may also lead to small changes in peak positions, which is disturbing. There are several possible

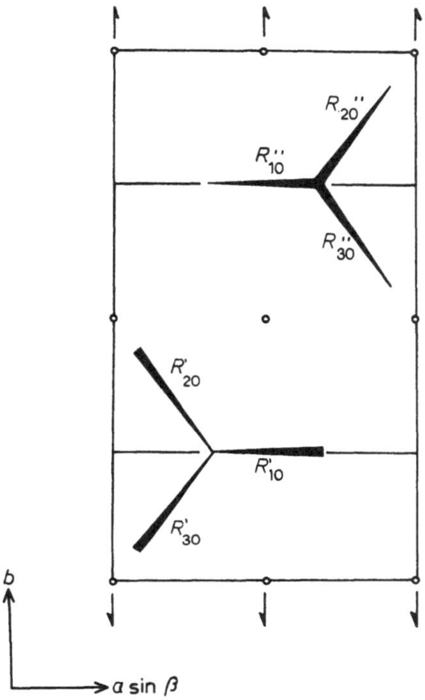

Fig. 4. The crystallographic unit cell of η^6-$C_6H_6Cr(CO)_3$[$Cr(CO)_3$ units only]

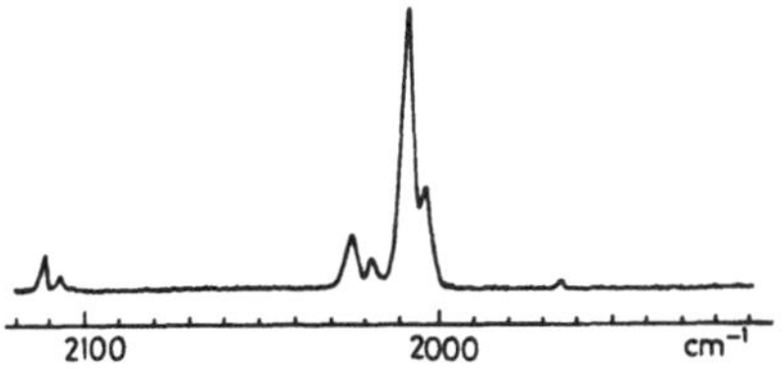

Fig. 5. Raman spectrum typical of $M(CO)_6$ (M = Cr,Mo,W) in the 2000 cm^{-1} region

explanations for this; interference phenomena, the Christiansen filter effect, a decrease in particle size or introduction of defects leading to a breakdown in the basis for a factor group analysis or the increased density of surface states being amongst the more probable. In one case, grinding has been reported as leading to a polymorphic change[64]; in such cases grinding under refrigeration is indicated. Except for ionic species nujul and similar mulls are to be avoided – for molecular species they tend to give a mixture of solution and crystal spectra.

For isolated molecules a variety of approaches have proved useful in the interpretation of vibrational spectra. Firstly, a species may approximate to a symmetry higher than its actual. In such cases a correlation with-descent in symmetry from – the higher symmetry usually simplifies the interpretation of its spectra. Secondly, local group vibrations, essentially uncoupled from the vibration of other equivalent or near-equivalent groups, may occur. Thirdly, chemically distinct groups may couple together. All of these features find a parallel in the vibrational spectra of crystalline metal carbonyls with consequent modification of the formal factor group predictions.

A good example of a higher-than-actual symmetry is provided by hexamethylbenzenetricarbonylchromium. In the crystal structure the threefold axes of the $Cr(CO)_3$ groups are almost parallel both to each other and to one of the (symmetry determined) crystallographic axes (Fig. 6)[65]. It follows that the dipole moment

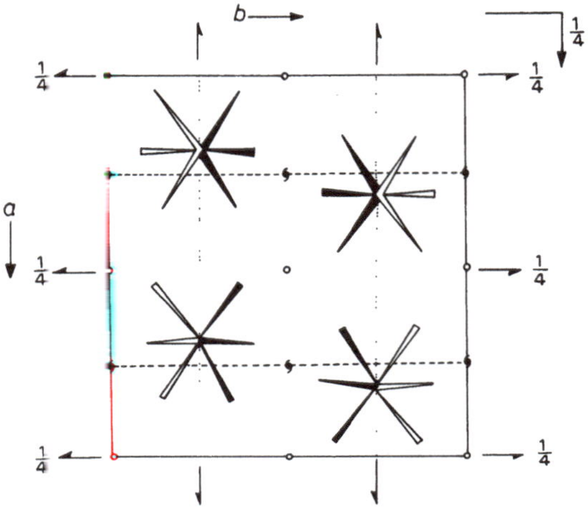

Fig. 6. The crystallographic unit cell of η^6-$C_6Me_6Cr(CO)_3$[$Cr(CO)_3$ units only]

change associated with the totally symmetric ν(CO) vibration of each $M(CO)_3$ unit is, effectively, along a crystallographic axis. As a consequence only one of the three formally infrared-active factor group components gives rise to a strong peak in the infrared spectrum[66]. Corresponding simplifications occur for the other ν(CO) infrared and Raman predictions[B10]. It is evident that a comparison between observed spectra and factor group predictions could, in principle, lead to crystal structure information (particularly if singly crystal data were available). This expectation has achieved some realization[67]; in the metal carbonyl field the most that has been achieved is a suggestion that a true unit cell is twice as large as that suggested by an X-ray analysis[B9].

In contrast, it has been suggested that the ν(CO) vibrational unit cell of $\eta^5 C_5H_5Fe(CO)_2 \cdot SnPh_3$ is only one half of the crystallographic to give a situation analogous to that of molecular group vibrations[68]. In the crystal structure of this compound the carbonyl groups lie in parallel sets of planes throughout the crystal (Fig. 7); there are at least two such planes in the crystallographic unit cell[69]. Vibrationally, the CO planes appear to be well insulated from each other and the infrared and Raman spectra can best be explained in terms of a unit cell which is one half of the crystallographic. The problem is a complicated one in that the unit cell contains two symmetry-distinct molecules; further reinvestigation has revealed errors in the frequencies earlier reported which make a frequency distinction between

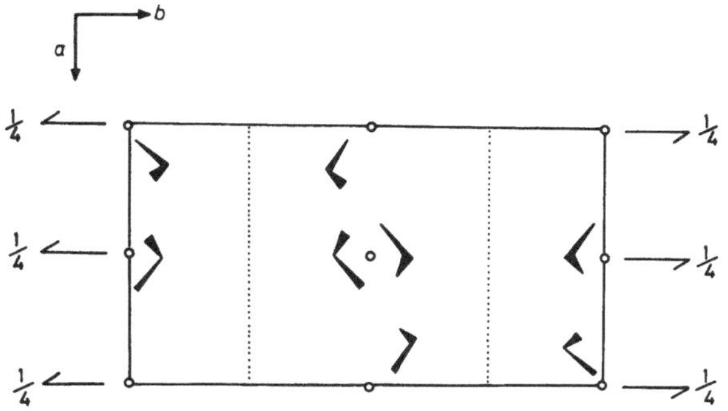

Fig. 7. The crystallographic unit cell of η^5-$C_5H_5Fe(CO)_2SnPh_3$[$Fe(CO)_2$ units only]

infrared and Raman ν(CO) peaks less certain. The isomorphous replacement method – *vide infra* – has, however, demonstrated that the conclusions reached earlier are correct[70].

We have seen that it is a characteristic of the fundamental vibrations of crystals studied by infrared or Raman spectroscopy that translationally-related molecules vibrate in phase. This is not to say that in the absence of electromagnetic radiation there would be such phase coherence. If proximate molecules are strongly coupled together we would expect that the 'natural' vibrations would involve a strong phase relationship (not necessarily all in-phase) between them. In contrast, in the limit of zero coupling there would be no phase relationship; each molecule would vibrate independently. In other words, strong factor group coupling will lead to vibrations delocalized over the lattice (in the limit, over the whole lattice). In this latter situation, what happens when one of the vibrators in replaced by a similar, but different, vibrator in a (random) mixed crystal? For instance, when ^{13}CO replaces ^{12}CO?

This, and related problems have been studied by in some detail by Kariuki and Kettle[71, 72]. They have defined three classes of behaviour; one mode, two mode and intermediate. Two mode behaviour occurs when the observed spectra are, essentially, a superposition of those of the (two) pure components. One mode behaviour occurs when a *single* spectral pattern, resembling that of either pure component, occurs at a frequency usually intermediate between those of the corresponding bands in the pure components. These two extremes of behaviour may be related to the extent of natural phase coherence of the vibrators concerned. Intermediate behaviour is believed to occur when there is only a limited extent of natural phase coherence. Qualitative application of perturbation theory indicates that the transition between these classes of behaviour depends on the magnitude of intermolecular vibrational coupling and zeroth-order (*i.e.* pure crystalline component) frequency separations as indicated in Fig. 8. A good example of intermediate and two mode behaviour is shown by the two $\nu(CO)A_1$ modes of $Mn(CO)_5Br$ when mixed either with the isomorphous compound $Re(CO)_5Br$ or with ^{13}CO-doped $Mn(CO)_5Br$. As shown in Fig. 9, the higher frequency (totally symmetric equatorial CO stretch vibration) displays two-mode behaviour. When mixed crystals containing $Re(CO)_5Br$ or $Mn(^{12}CO)_n(^{13}CO)_{5-n}Br$ in ^{13}CO $Mn(CO)_5Br$ are studied in the Raman (for the reasons indicated above the Raman spectra are easier to study, although similar

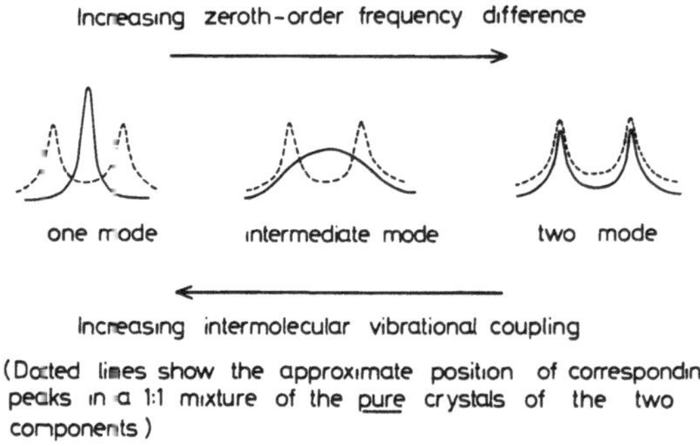

Fig. 8. Spectroscopic characteristics of one, intermediate and two mode behaviour

phenomana occur, but less clearly, in the infrared), each species present gives rise to its own characteristic frequency (static field effects seem to be small). Thus, as ^{13}CO enrichment is increased new peaks originating in the variety of di-substituted species are observed. These observations provide an explanation for the 'missing' isotopic peaks reported by Butler and Shaw[A79].

The lower frequency, axial CO stretching mode, peak is well separated in the manganese and rhenium compounds. As suggested by Fig. 8 this is a situation in which intermediate mode behaviour may be expected. Figure 10 shows that this is

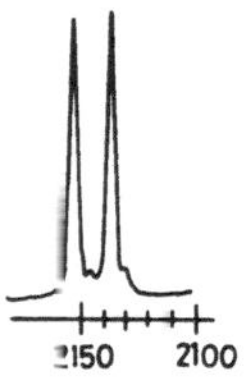

Fig. 9. The Raman spectrum of the A_1(radial)-derived ν(CO) modes in crystalline $Mn_xRe_{1-x}(CO)_5Br$

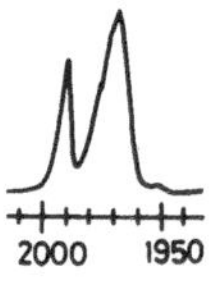

Fig. 10. The Raman spectrum of the A_1(axial)-derived ν(CO) modes in crystalline $Mn_xRe_{1-x}(CO)_5Br$

so, the peaks shown in the figure on the broad intermediate mode band probably reflecting the high statistical probability of occurrence of some patterns of aggregation of the two components. Increasing the zeroth order frequency separation between $Mn(CO)_5Br$ and the second component by ^{13}CO substitution leads, as expected from Fig. 8, to two mode behaviour (Fig. 11). The other Raman-active modes in $Mn(CO)_5Br$ display one-mode behaviour, but this phenomenon is more clearly shown by the E_g-derived modes of $M(CO)_6$, M = Cr, Mo, W. Figure 12 shows this for a 1 : 1 mixture of $Cr(CO)_6$ and $W(CO)_6$. Also included in this figure is the A_{1g}-derived feature, which displays two mode behaviour.

Six comments are appropriate at this point. Firstly, it is the experience of the reviewer that chemically-similar compounds with very similar infrared and Raman spectra in factor-group split regions are isomorphous. This method is probably at least as reliable as X-ray powder methods when there is a significant change in scattering factors between the two compounds studied (*e.g.* bromo-derivatives of

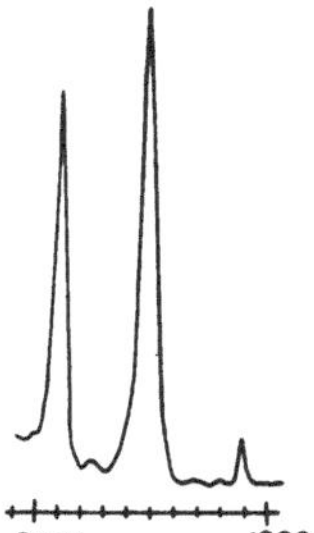

Fig. 11. The Raman spectrum of the a_1(axial)-derived ν(CO) modes in crystalline $Mn(^{12}CO)_{5-x}(^{13}CO)_xBr$

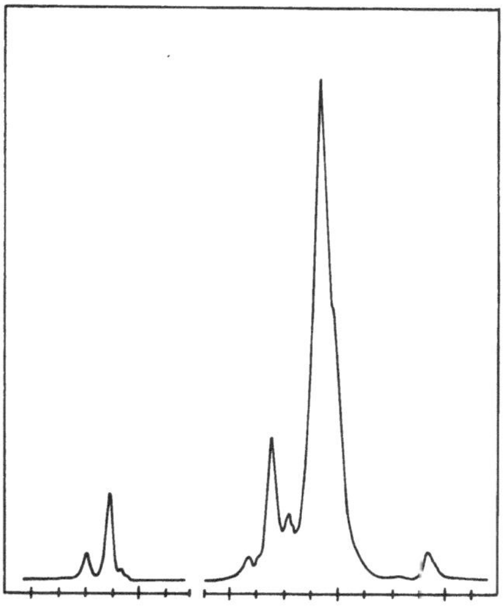

Fig. 12. The Raman spectrum of 1 : 1 co-sublimed mixture of $Cr(CO)_6$ and $W(CO)_6$ in the 2000 cm^{-1} region. The A_{1g} mode is at high frequency

first and third row transition elements). Secondly, there is a clear moral to be learnt from the mixed crystal studies. A factor group approach may be necessary to explain crystalline spectral features derived from one molecular internal mode involving a particular symmetry coordinate but not for others. Thus, the factor-group method is needed to explain the E_g-derived ν(CO) modes features in the Raman spectra of $M(CO)_6$, M = Cr, Mo, W, but not for the A_{1g}-derived ν(CO) mode features. Thirdly, it is clear that in the ν(CO) region impurities may be manifest not by additional peaks but by changes in band position, breadth or intensity. This will probably be particularly true for infrared spectra because of the (long range) dipole-dipole coupling which presumably occurs for most strong bands. An observation made by most workers in the field may thus be explained; the observation that the mull or disc infrared spectrum of a carbonyl does not agree with the literature value, the disagreement being outside the aggregate instrumental error and, in any case, not constant throughout all the peaks in the ν(CO) region. Fourthly, the mixed crystal data indicate that differences in molecular symmetry and consequent differences in molecular normal modes (such as occur on ^{13}CO substitution) may be swamped by intermolecular vibrational coupling (hexacarbonyls in the E_g-derived ν(CO) region show a slightly broadened one-mode behaviour on ^{13}CO substitution). The differences in site-orientation which occur for a labelled molecule may similarly be swamped. Fifthly, these observations strongly suggest that, when more than one set of crystallographically-related molecules occupy a unit cell, it is probably valid to treat these different molecules as identical in a factor group analysis [η^5-$C_5H_5Fe(CO)_2$-$SnPh_3$ provides an example of this[68, 69]]. Sixthly, these studies have revealed that there is no necessary 1 : 1 correspondence between modes arising from $M(^{12}CO)_n$ and $M(^{12}CO)_{n-1}$ ^{13}CO units in crystal spectra. Thus a new explanation has to be sought for some peaks previously assigned as isotopic (incidentally, a test which has sometimes been available but which has not been used is that, for instance, any natural abundance $M(^{12}CO)_5(^{13}CO)$ peaks in the spectrum of $M(^{12}CO)_6$ will have no counterparts in the spectrum of near-completely substituted $M(^{13}CO)_6$). Perhaps the most probably explanation of such peaks is that they are multiphonon in origin, strengthened by mixing with, and stealing intensity from, a nearby active fundamental mode of the same symmetry species[18]. Single crystal studies, which for factor group components give information on the symmetry species of the mode associated with a particular vibrational feature, seem to be in general support of this hypothesis.

Granted that CO stretching vibrations may couple intermolecularly in a crystal, the question of the mechanism involved arises. The simplest analysis would be one based on a multi-polar expansion about each molecular centre[76, 77]. The only paper on this subject is that of Bullitt and Cotton who showed that a dipole-dipole coupling model provided an explanation. Unfortunately, their model was parameterized and there is no independent check on the values of the best-fit parameters which they deduced. In principle a crystal structure is needed (to give angles and distances between dipoles) together with infrared intensity data (to give a measure of the dipole strength). Even with such data, which lead to a non-parameteried treatment, there is some uncertainty – how applicable are solution intensities to the crystal state (solid state intensities are difficult to determine), precisely where are the dipoles to be placed and how oriented? Burrows and Kettle[73] have shown that the predictions obtained with such a model are not good and that some additional coupling mechanism is probably involved. There is evidence that this is the case. The factor group splitting of the ν(CO) E_g-derived modes of metal hexacarbonyls almost certainly requires a quadrupole-quadrupole mechanism (there is negligible infrared activity). Further, the splittings are entirely comparable to those that are observed in infrared-active features. In high CO concentrations it seems, then, that both dipole-dipole and quadrupole-quadrupole coupling must be considered (there is no evidence for dipole-quadrupole coupling). The absence of factor group splitting on the high frequency A_1 ν(CO) mode of $Mn(CO)_5Br$ and related species indicates that octupole-octupole terms are probably negligible; similarly the two-mode behaviour of the $\nu(CO)A_{1g}$ – derived feature of hexacarbonyls discounts, not surprisingly, the involvement of hexadecapole-hexadecapole interactions.

As indicated above, an area of future development is the exploration of the $k \neq 0$ (multiphonon) aspects of vibrational spectra; as the techniques of normal coordinate analyses are developed it should become increasingly possible to explore, at least qualitatively, the whole of the Brillouin zone and to discover whether, for instance the dispersion on internal modes is negligible (as is usually assumed with rather little justification). One aspect which has been studied is that of L.O./T.O. splitting a phenomenon by which that which is, on a factor group analysis, predicted to be a single peak is, in fact, two. Such splittings can be large[74], arising from the difference in (assumed dipole-dipole) coupling energies with direction of propagation of the vibration in the crystal. The phenomenon is most readily observed in single crystal Raman studies and so simultaneous infrared and Raman activity is needed. That is, centrosymmetric crystals are excluded from study. A carbonyl to be studied must form large crystals (ca. 10 mm^3 minimum), be little coloured and be air and laser beam stable and have simple and unambiguous factor group predictions (*i.e.* preferably only one type of crystallographically-distinct CO group). There are a few metal carbonyls which satisfy these conditions but despite a rather thorough study, L.O. – T.O. splitting has yet to be observed in the ν(CO) region[75)].

In this section of this review there has been scant attention paid to singly crystal studies; this is because it is usually difficult to generalize the lessons from a particular study – such generalizations as are possible have already been made. It would not be appropriate, however, to conclude this review without recognizing that much of our understanding of crystal spectra comes from singly crystal studies. Although Raman studies are not easy, single-crystal infrared studies in the ν(CO) are yet more difficult. In order that a crystal be not completely opaque in this spectral region it is necessary to work with extremely thin but large crystals. It is therefore encouraging to conclude by noting that data from such studies are becoming available[B3, B4, B6, B7, B9].

Acknowledgement. The author is indebted to D. F. Shriver and G. Bor for helpful comments.

5. References

1) Braterman, P. S.: Metal carbonyl spectra. London: Academic Press 1975
2) Cotton, F. A., Kraihanzel, C. S.: J. Amer. Chem. Soc. *84*, 4432 (1962)
3) Kraihanzel, C. S., Cotton, F. A.: Inorg. Chem. *2*, 533 (1963)
4) Cotton, F. A., Liehr, A. D., Wilkinson, G.: J. Inorg. and Nucl. Chem. *2*, 175 (1955)
5) Wilson, W. E.: Z. Naturforsch. *136*, 349 (1958)
6) Cable, J. W., Sheline, R. K.: Chem. Rev. *56*, 1 (1956)
7) Kaesz, H. D., Bau, R., Hendrickson, D., Smith, J. M.: J. Amer. Chem. *Soc. 89*, 2844 (1967)
8) Davidson, A., Faller, J. W.: Inorg. Chem. *6*, 845 (1967)
9) Abel, E. W., Wilkinson, G.: J. Chem. Soc. *1959*, 150
10) El-Sayed, M. A., Kaesz, H. D.: J. Mol. Spec. *9*, 310 (1962)
11) Orgel, L. E.: Inorg. Chem. *1*, 25 (1962)
12) Poilblanc, R., Bigorgne, M.: Bull. Soc. Chem. France *1962*, 1303
13) Miller, J. R.: J. Chem. Soc. (*A*), 1885 (1971)
14) Kaesz, H. D., Bau, R., Hendrickson, D., Smith, J. M.: J. Amer. Chem. *89*, 2844 (1967)
15) Ottesen, D. K., Gray, H. B., Jones, L. H., Goldblatt, M.: Inorg. Chem. *12*, 1051 (1973)
16) Clark, R. J. H., Crosse, B. C.: J. Chem. Soc. (*A*), 224 (1969)
17) Fenske, R. F., DeKoch, R. L.: Inorg. Chem. *9*, 1053 (1970)
18) Miller, J. R.: Inorg. Chem. Acta *2*, 421 (1968)
19) Kettle, S. F. A., Paul, I.: Inorg. Chim. Acta *2*, 15 (1968)
20) Edgell, W. F.: J. Chem. Phys. *75*, 1343 (1971)
21) Representation theory of the symmetric group Robinson, B. (ed.). Toronto: University of Toronto Press 1961
22) Bor, G.: J. Organometal Chem. *65*, 81 (1974)
23) Sbrignadello, G., Bor, G., Maresca, L.: J. Organometal Chem. *46*, 345 (1972)
24) Sbrignadello, G., Tomat, G., Magon, L., Bor, G.: Inorg. Nucl. Chem. *Letts. 9*, 1073 (1973)
25) Watson, J. K. G.: Can. J. Physics *43*, 1996 (1965)
26) Dalton, B. J.: Mol. Phys. *11*, 265 (1966)
27) Stammreich, H., Kawai, K., Tavero, Y., Behmoiras, J., Bril, A.: J. Chem. Phys. *32*, 1452 (1960)
28) Edgell, W. F., Lyford(IV), J.: J. Chem. Phys. *52*, 4329 (1970)
29) Terzis, A., Spiro, T. G.: Inorg. Chem. *10*, 643 (1971)
30) Long II, T. V., Plane, R. A.: J. Chem. Phys. *43*, 457 (1965)
31) Chantry, G. W., Plane, R. A.: J. Chem. Phys. *35*, 1027 (1961)
32) Hester, R. E.: in: Raman spectroscopy. Szymanski, H. A. (ed.). New York: Plenum Press 1967
33) Kettle, S. F. A., Paul, I., Stamper, P. J.: Chem. Comms. *1970*, 1724
34) Kettle, S. F. A., Paul, I., Stamper, P. J.: Chem. Comms. *1971*, 235
35) Wolkenstein, M.: Dokl. Alcad Naulc SSSR *32*, 185 (1941)
36) Eliashevich, M., Wolkenstein, M.: Fiz. Zh *9* 101 (1945)
37) Eliashevich, M., Wolkenstein, M.: Fiz. Zh. *9*, 326 (1945)
38) Lucknar, N., Kettle, S. F. A.: J. Chem. Phys., in press
39) Cotton, F. A., Musco, A., Yagupsky, G.: Inorg. Chem. *6*, 1357 (1967)
40) Manning, A. R., Miller, J. R.: J. Chem. Soc. *A*, 1521 (1966)
41) Halford, R. S.: J. Chem. Phys. *14*, 8 (1946)
42) Adams, D. M., Newton, D. C.: Tables for factor group and point group analysis (Beckman-RIIC Limited)
43) Fateley, W. G., Dollish, F. R., McDeritt, N. T., Bentley, F. F.: Infrared and Raman selection rules for molecular and lattice vibration: The correlation method. New York: Wiley-Interscience 1972
Another particularly useful book is
Infrared and Raman spectra of crystals. Turell, G. (ed.). London: Academic Press 1972
44) Horning, D. F.: J. Chem. Phys. *16*, 1063 (1945)

45) Dows, D. A.: J. Chem. Phys. *63*, 168 (1966)
46) Winston, H.: J. Chem. Phys. *19*, 156 (1951)
47) Bertie, J. E., Bell, J. W.: J. Chem. Phys. *54*, 160 (1971)
48) Bertie, J. E., Kopelman, R.: J. Chem. Phys. *55*, 3613 (1971)
49) Kopelman, R.: J. Chem. Phys. *47*, (1967)
50) DeAngelis, B. A., Newnham, R. E., White, W. B.: Amer. Mineral. *57*, 255 (1972)
51) Warren, J. L.: Rev. Mod. Phys. *40*, 38 (1968)
52) Chen, S. H., Drorak, V.: J. Chem. Phys. *48*, 4060 (1958)
53) Maradudin, A. A., Vosko, S. H.: Rev. Mod. Phys. *40*, 1 (1968)
54) Walnut, T. H.: J. Chem. Phys. *31*, 361 (1959)
55) Donnay, J. D. H., Turrell, G.: Chem. Phys. *6*, 1 (1974)
56) Haas, C.: Spect. Acta *8*, 19 (1956)
57) Hexter, R. M.: J. Chem. Phys. *33*, (1960)
58) Beattie, I. R., Cheetham, N., Gardner, M., Rogers, D. E.: J. Chem. Soc (*A*), 2240 (1971)
59) Pawley, G. S., Cyvin, S. J.: J. Chem. Phys. *52*, 4073 (1970)
60) De Hosson, J. T. M.: Comp. Phys. Comm. *10*, 194 (1975)
61) Arif, M., Kettle, S. F. A.: unpublished observations
62) Buttery, H. J., Keeling, G., Kettle, S. F. A., Paul, I., Stamper, P. J.: J. Chem. Soc. (*A*), 2077 (1969)
63) Adams, D. M., Squire, A.: J. Chem. Soc. (*A*), 814 (1970)
64) Buttery, H. J., Keeling, G., Kettle, S. F. A., Paul, I., Stamper, P. J.: Disc. For. Soc. *47*, 48 (1969)
65) Bailey, M. F., Dahl, L. F.: Inorg. Chem. *4*, 1298 (1965)
66) Buttery, H. J., Keeling, G., Kettle, S. F. A., Paul, I., Stamper, P. J.: J. Chem. Soc. (*A*), 2224 (1969)
67) See, for instance, Scott, J. F. in: Vibrational spectra and structure. Vol. 5. Dung, J. R. (ed.). Amsterdam: Elsener 1976
68) Buttery, H. J., Kettle, S. F. A., Keeling, G., Paul, I., Stamper, P. J.: J. Chem. Soc. (Dalton) *1972*, 2487
69) Bryan, R. F.: J. Chem. Soc. (*A*), 192 (1967)
70) Barker, S. L., Harland, L., Kettle, S. F. A.: to be published
71) Kariuki, D. N., Kettle, S. F. A.: J. Organometal Chem. *105*, 209 (1976)
72) Kariuki, D. N., Kettle, S. F. A.: Inorg. Chem., in press
73) Burrows, E., Kettle, S. F. A.: unpublished results
74) Arguello, C. A., Rousseau, D. L., Porto, S. P. S.: Phys. Rev. *181*, 1351 (1969)
75) Kariuki, D. N., Kettle, S. F. A.: unpublished observations
76) Frech, R., Decius, J. C.: J. Chem. Phys. *54*, 2374 (1971)
77) Carlson, R. E., Decius, J. C.: J. Chem. Phys. *58*, 4919 (1973)
78) Johnson, B. F. G., Lewis, J., Miller, J. R., Robinson, B. H., Robinson, P. W., Wojcicki, A.: J. Chem. Soc. (*A*), 522 (1968)
79) A further approach to this problem has recently been given by Bigorgne, M.: Spect. Acta, *32A*, 1365 (1976)

Received March 15, 1977

Table A

Ref.		
A1	Hsieh, A. T. T., Mays, M. J., Platt, R. H.: J. Chem. Soc. (A) *1971*, 3296	Infrared (and Mössbauer) of $CdFe(CO)_4$ and $L_2MFe(CO)_4$, M = Zn or Cd; L = N donor
A2	Chatt, J., Leigh, G. J., Thankarajan, N.: J. Chem. Soc. (A) *1971*, 3168	ν(CO) in crystalline $[Ru(NH_3)5(CO)]^{2+}$ increases with counter-anion radius (also related cations reported)
A3	Cataliotti, R., Paliani, G., Poletti, A.: Chem. Phys. Lett. *11*, 58 (1971)	IR spectrum of crystalline and liquid $Co(CO)_3NO$ as a function of temperature
A4	Bigorgne, M., Tripathi, J. B. P.: J. Mol. Struct. *10*, 449 (1971)	Raman and IR of liquid and solid $(CO)Fe(PF_3)_4$
A5	Keiter, R. L., Shah, D. P.: Inorg. Chem. *11*, 191 (1972)	Effect of a positive charge on ν(CO) in tungsten carbonyl complexes involving a chelating phosphine
A6	Davidson, G., Andrews, D. C.: J. Chem. Soc. (Dalton) *1972*, 126	Assignment of IR and Raman spectra of η^3-$C_3H_5Mn(CO)_4$
A7	Andrews, D. C., Davidson, G.: J. Organomet. Chem. *35*, 161 (1972)	Assignment of IR and Raman spectra of $C_2H_4Fe(CO)_4$. Includes crystal data
A8	Braunstein, P., Dehand, J.: Chem. Comms. *1972*, 164	Metal-metal IR frequencies etc. in a variety of metal carbonyl species
A9	Ziegler, R. J., Burlitch, J. M., Hayes, S. E., Risen, W. M.: Inorg. Chem. *11*, 702 (1972)	Assignment and vibrational analysis of infrared and Raman spectra of $M[Co(CO)_4]_2$ M = Zn, Cd, Hg. Uses isolated molecule approximation
A10	Crease, A. E., Legzdino, P.: Chem. Comm. *1972*, 268	ν(CO) infrared spectra changes indicate coordination of lanthanides to the CO oxygen
A11	Benedetti, E., Braea, G., Sbrana, S., Sabretti, F., Grassi, B.: J. Organomet. Chem. *37*, 361 (1972)	Infrared and assignment of $[Ru(CO)_3X_2]_2$ X = Cl, Br
A12	Edgell, W. F.: Ions Ion Pair Org. Reat *1*, 153 (1972)	Ion pair vibrational spectra of $Co(CO)_4^-$ salts
A12(a)	Barna, G., Butler, I. S.: Can. J. Spect. *17*, 2 (1972)	Vapour phase and solution infrared spectra of $Mn(CO)_4(NO)$
A13	Schaefer, L., Begun, G. M., Cyvin, S. J.: Spec. Acta. *28 (A)*, 803 (1972)	Raman spectra of solid and solution η^6-$C_6H_6Cr(CO)_3$
A14	Andrews, D. C., Davidson, G.: J. Chem. Soc. (Dalton) *1972*, 1381	Assignment of IR and Raman of η^3-$C_3H_5Co(CO)_3$

Table A (continued)

Ref.		
A15	Clarke, H. L., Fitzpatrick, N. J.: J. Organomet. Chem. *40,* 379 (1972)	Vibrational spectra and analysis of η^1-$C_3H_5Mn(CO)_5$
A16	Duddell, D. A., Kettle, S. F. A., Kontnik-Matecka, B. T.: Spect. Acta *28A,* 1571 (1972)	Raman and IR of η^4-$C_4H_6Fe(CO)_3$ in a variety of phases
A17	Van Bronswyk, W., Clark, R. J. H.: Spect. Acta *28A,* 1429 (1972)	Raman, infrared and vibrational analysis of $Os_4O_4(CO)_{12}$
A18	Dobson, G. R., Brown, R. A.: Jour. Inorg. Nucl. Chem. *34,* 2785 (1972)	Low frequency IR of several *cis*(bidentate) $M(CO)_4$ species and correlation with reactivity
A 19	Hyams, I. J., Lippincott, E. R.: Spect. Acta *28A,* 1741 (1972)	Raman and infrared of η^6-$C_6H_6Cr(CO)_3$ and η^6-$C_6D_6Cr(CO)_3$ and their analysis
A20	Jerrigan, R. T., Brown, R. A., Dobson, G. R.: J. Coord. Chem. *2,* 47 (1972)	Spectra and analysis of ν(CO) in a variety of (bidentate)$M(CO)_4$ complexes, M = Cr, Mo, W
A21	Quinby, M. S., Feltham, R. D.: Inorg. Chem. *11,* 2468 (1972)	The infrared spectra and their analysis for a variety of complexes containing a Ru-A-B unit (AB = CO, NO, N_2)
A22	Clarke, H. L., Fitzpatrick, N. J.: J. Organometal. Chem. *43,* 405 (1972)	Spectra and vibrational analysis of a variety of allyl $Co(CO)_3$ complexes compared with a simple M.O. model
A23	Labrone, D., Poilblanc, R.: Inorg. Chim. Acta *6,* 387 (1972)	Infrared study and assignment of ν(CO) in several $Co_4(CO)_{12-x}L_x$ species (L = P ligand)
A24	Onaka, S.: Nippon Kag. Kaishi *1972,* 1978	Raman and IR of I_3Sn-$Mn(CO)_5$
A25	Davidson, G., Duce, D. A.: J. Organometal. Chem. *44,* 365 (1972)	Vibrational spectra and assignments for η^4-$(C_4H_6)_2Fe(CO)$
A26	Cleare, M. J., Fritz, H. P., Griffith, W. P.: Spect. Acta *28A,* 2019 (1972)	Raman, IR and assignments for $[M(CO)X_5]^{2-}$ anions, M = Os, Ru, Ir, Rh
A27	Lokshin, B. V., Klemenkova, Z. S.: Spect. Acta *28A,* 2209 (1972)	Raman, IR and analysis of dissolved and crystalline η^5-$C_5H_5Re(CO)_3$
A28	Wozniak, W. T., Sheline, R. K.: J. Inorg. Nucl. Chem. *34,* 3765 (1972)	The preparation, IR (solution), Raman (crystal) and analyses in the ν(CO) region of ^{13}CO-enriched $M_2(CO)_{10}$, M = Mn, Re
A29	Pomerry, R. K., Gay, R. S., Evans, G. O., Graham, W. A. G.: J. Amer. Chem. Soc. *94,* 272 (1972)	^{13}CO exchange in *cis* $Ru(CO)_4(SiCl_3)_2$ occurs only for the two equatorial CO groups; followed by ν(CO) IR spectroscopy

Table A (continued)

Ref.		
A30	Lokoshin, B. V., Klemenkora, Z. S., Makorov, Y. V.: Spect. Acta *28A*, 2209 (1972)	Raman and IR (of solution and crystal) of η^5-$C_5H_5Re(CO)_3$
A31	Kristoff, J. S., Shriver, D. F.: Inorg. Chem. *12*, 1788 (1973)	The vibrational spectra [especially ν(CO) asymm] of η^5-$C_5H_5Fe(CO)_2CN$ coordinated through the CN to a variety of B, Al or Ga-containing electron acceptors
A32	Whyman, R.: J. Organomet. Chem. *63*, 467 (1973)	Infrared evidence for the reversible formation of $M(CO)_3P\phi_3$, M = Pd, Pd under high CO pressure
A33	Lokshin, B. V., Pasinsky, A. A., Kolobova, N. E., Anisimov, K. N., Makarov, Y. V.: J. Organomet. Chem., *55*, 315 (1973)	A study of the protonation of η^5-$C_5H_5M(CO)_4$ (M = V,Nb), η^5-$C_5H_5Re(CO)_3$ and their phosphine derivatives by following ν(CO)
A34	Whyman, J.: J. Organometal. Chem. *56*, 339 (1973)	Infrared evidence for the reversible formation of $H_2Ru(CO)_3PPh_3$ under high H_2 pressure
A35	Morris, D. E., Tinker, H. B.: J. Organometal. Chem. *49*, C53 (1973)	Infrared data [ν(CO)] indicates that under high CO pressure $[Rh(CO)_2X]_2$ is in equilibrium with $Rh(CO)_3X$
A36	Kristoff, J. S., Nelson, N. J., Shriver, D. F.: J. Organometal. Chem. *49*, C82 (1973)	Infrared spectra [ν(CO)] of $Co_2(CO)_8$. $AlBr_3$ and related compounds indicate that the Lewis acid interacts with a bridging CO group
A37	Kalbfus, W., Kiefer, J., Schwarzhauss, K. E.: Z. Naturforsch. *28B*, 503 (1973)	The preparation and IR spectrum of $GaCo_3(CO)_{12}$
A38	Ottesen, D. K., Gray, H. B., Jones, L. H.: Inorg. Chem. *12*, 1051 (1973)	Raman and IR of ^{13}C and ^{18}O substituted $Mn(CO)_5Br$ together with a detailed analysis
A39	Jones, L. H., McDowell, R. S., Swanson, B. I.: J. Chem. Phys. *58*, 3757 (1973)	Raman and IR spectra for ^{13}C, ^{18}O and ^{15}N substituted $Co(CO)_3NO$ together with a detailed analysis
A40	Butler, I. S., Fenster, A. E.: J. Organmetal. Chem. *51*, 307 (1973)	The IR spectra and their analysis of a variety of ^{13}CO substituted η^5-$C_5H_5M(CO)_n$ carbonyls and their derivatives (including CS) M = Mn, V, Co
A41	Davidson, G., Riley, E. M.: J. Organometal. Chem. *51*, 297 (1973)	The Raman and IR spectra of η^5-$C_5H_5W(CH_3)(CO)_3$ and their analysis

Table A (continued)

Ref.		
A42	Adams, D. M., Fernando, W. S.: Inorg. Chim. Acta *7*, 277 (1973)	Vibrational spectra and assignments of $Cr(CO)_4N$, N = 2,2,1-bicycloheptadiene
A43	Chenskaya, T. B., Lokshin, B. V., Kritskaya, I. I.: Izv. Akad. Nauk SSSR *1973*, 1146	Infrared spectra and their analysis for a variety of allyl $Fe(CO)_3X$ species in the ν(CO) region
A44	Kubas, G. J., Spiro, T. G.: Inorg. Chem. *12*, 1797 (1973)	Resonance Raman spectrum and analysis of $(\mu\text{-}A)_2Fe_2(CO)_4$ A = ditertbutylacetylene
A45	Burnham, R. A., Stobart, S. R.: J. Chem. Soc. (Dalton) *1973*,	Raman, IR (and mass spectra) of $(CH_3)_3MMn(CO)_5$, M = Si, Ge, Sn
A46	Colton, R., Garrard, J. E.: Aust. J. Chem. *26*, 1781 (1973)	Infrared spectra of $[Re(CO)_3X]_3$, X = Cl, Br, I
A47	Klassen, K. L., Duffy, N. V.: J. Inorg. Nucl. Chem. *35*, 2602 (1973)	Infrared spectra of *cis* and *trans* $PtCl_2(CO)PR_3$ (R = aromatic)
A48	Onaka, S.: Bull. Chem. Soc. Jap. *46*, 2444 (1973)	Infrared spectra and analysis of $X_3MMn(CO)_5$ M = Si, Ge, Sn
A49	San Filippo, Jr., J, Snaidoch, H. J.: Inorg. Chem. *12*, 2326 (1973)	Low frequency Raman spectra of metal-metal bonded system [includes $Fe_2(CO)_9$ with ν(CO)], $Fe_3(CO)_{12}$, $[\eta^5\text{-}C_5H_5Fe(CO)_2]_2$ and $[\eta^5\text{-}C_5H_5Fe(CO)]_4$
A50	Van de Berg, G. C., Oskam, A., Vrieze, K.: J. Organometal. Chem. *57*, 329 (1973)	Raman, IR and assignments of $X_3MCo(CO)_4$, M = Si, Ge, Sn
A51	Butler, I. S., Barna, G. G.: J. Raman Spec. *1*, 141 (1973)	The Raman and IR spectra and their assignment for $LM(CO)_4$, M = Cr, Mo, W
A52	Oetker, C. J., Beck, W.: Spect. Acta *29A*, 1975 (1973)	Resonance Raman study of $W(CO)_5L$, L = I^- phthalimide, succinimide and saccharide complexes (and also azides of Pt, Pd, Au and U)
A53	Adams, D. M., Squire, A.: J. Organometal. Chem. *63*, 381 (1973)	The vibrational spectra and their assignment for $\eta^5\text{-}C_5H_4RMn(CO)_3$, (R = H, Me)
A54	King, R. B., Saran, M. S.: Inorg. Chem. *13*, 74 (1974)	The preparation, vibrational spectra and analysis of compounds of the general form (tert $BuNC)_nM(CO)_{6-n}$ M = Cr, Mo, W
A55	Butler, I. S., Sawai, T.: Inorg. Chem. *12*, 1994 (1973)	Infrared of $\eta^5\text{-}C_5H_5Mn(CO)_2S$, S = sulphur ligand demonstrate Mn-S conformational isomerism. Isotopic data are included

Table A (continued)

Ref.		
A56	Bor, G., Sbrignadello, G., Natile, G.: J. Organometal. Chem. *56,* 357 (1973)	Infrared and analysis of $Co_3(CO)_4(SMe)_5$; uses ^{13}CO data
A57	Delbeke, F. T., Van der Kelen, G. P.: J. Organometal. Chem. *64,* 239 (1974)	Synthesis and IR spectra of $LCr(CO)_5$ species (L = P, As, Sb ligand). Includes ^{13}CO data
A58	Scorell, W. M., Spiro, T. G.: Inorg. Chem. *13,* 304 (1974)	Raman and IR of $L_2Fe_2(CO)_6$, L = S ligand. Vibrational analysis and discussion of Raman intensities
A59	McLean, R. A. N.: Can. J. Chem. *52,* 213 (1974)	Raman and IR spectra of $[Mn(CO)_6][PF_6]$
A60	Onaka, S.: Nippon Kag. Kaishi *1974,* 255	The IR spectra and their analysis for $LMn(CO)_5$ species [L = GeH_3, $Ge(CH_3)_3$, SiH_3]
A61	Kilner, M.: Nature Phys. Sci. *1972,* 239	Infrared evidence for the species $Pt/Pd(CO)_3PPh_3$ and $Pd(CO)_2(PPh_3)_2$
A62	Parker, D. J.: J. Chem. Soc. (Dalton) *1974,* 155	Raman and infrared spectra in several phases for η^5-$C_5H_5Mn(CO)_3$ and η^5-$C_5D_5Mn(CO)_3$
A63	Young, F. R., Levenson, R. A., Memoring, M. N., Dobson, G. R.: Inorg. Chem. Acta *8,* 61 (1974)	Raman spectra for solutions of $LM(CO)_5$, M = Cr, W, L = Lewis base, below 680 cm^{-1}
A64	Rehder, R., Schmidt, J.: J. Inorg. Nucl. Chem. *36,* 333 (1974)	Infrared (and ^{51}V NMR) study of $[V(CO)_5L]^-$ L = P, As, Sb ligand
A65	Sellmann, D., Brandl, A., Endell, R.: Angew. Chem. *85,* 1122 (1973)	Infrared spectra of $N_2H_2[Cr(CO)_5]_2$ together with those of isotopic substituents on the N_2H_2 group
A66	Goggin, P. L., Mink, J.: J. Chem. Soc. (Dalton) *1974,* 534	The preparation and IR study of salts of $[Pd_2X_4(CO)_2]^{2-}$, $[Pd_4Cl_6(CO)_4]^{2-}$ and of $[PdX(CO)]_n$ and $Pd_6Cl_6(CO)_6(PhCN)_2$, X = Cl, Br
A67	Adams, D. M., Squire, A.: J. Chem. Soc. (Dalton) *1974,* 558	Raman and IR spectra and assignments for monosubstituted η^6-arene $Cr(CO)_3$ species including crystal data
A68	Noeth, H., Deberitz, J.: Kem. Kozlem. *40,* 9 (1973)	Infrared (and NMR) data on a variety of $M(CO)_n$-containing compounds involving phenyl-substituted heterocyclic aromatics as n- and π-ligands M = Cr, Mo, W, n = 3,4,5

Table A (continued)

Ref.		
A69	Palyi, G., Vizi-Orosz, A., Marko, L., Marcati, F., Bor, G.: J. Organometal. Chem. *66*, 295 (1974)	Infrared evidence for mixed-ligand bridged species $Rh_2(CO)_4XY$
A70	Lokshin, B. V., Makarov, Y. V., Klemenkova, Z. S., Kolobova, N. E., Anisimov, K. N.: Izv. Akad. Nauk SSSR *1974*, 710	Low frequency vibrational data for η^5-$RC_5H_4M(CO)_3$, M = Re, Mn
A71	Terzis, A., Strekas, T. C., Spiro, T. G.: Inorg. Chem. *13*, 1346 (1974)	Raman data (including intensities) and their interpretation for the low frequency region of $LM(CO)_5$, L = Sn, Ge ligand, M = Mn, Re
A72	Butler, W. M., McAllister, W. A., Risen, Jr., W. M.: Inorg. Chem. *13*, 1702 (1974)	The IR and Raman spectra (of solutions and crystals) and their analysis for $LFe(CO)_4^-$ salts L = $GeCl_3$, $SnCl_3$, $SnBr_3$ and $X_3MCo(CO)_3$ M = Ge, Sn, X = Cl, Br, I
A73	Bee, M. W., Kettle, S. F. A., Powell, D. B.: Spect. Acta *30A*, 585 (1974)	ν(CO) and far IR and Raman for $[M(NH_3)_5CO]^{2+}$ salts, M = Ru, Os and their interpretation (also the corresponding N_2 compounds)
A74	Jones, A. G., Powell, D. B.: Spect. Acta *30A*, 563 (1974)	Solution and crystalline infrared and Raman data for $NiL(CO)_3$, L = PPh_3, $AsPh_3$ and their assignment
A75	Andrews, D. C., Davidson, G.: H. Organometal. Chem. *74*, 441	Solution and crystal IR and Raman data for (η^2-maleic anhydride)$Fe(CO)_4$ and their assignment
A76	Darensbourg, D. J., Nelson, H. H., Hyde, C. L.: Inorg. Chem. *13*, 2135 (1974)	The IR spectra, ν(CO) absolute intensities and their analysis (also using data from ^{13}CO enriched species) for a variety of $LFe(CO)_4$ $L_2Fe(CO)_3$ species
A77	Braunstein, P., Dehand, J.: J. Organometal. Chem. *81*, 123 (1974)	Preparation and IR spectra of $ML_2[Mn(CO)_5]_2$ species M = Pd, Pt; L = a pyridine ligand
A78	Lokshin, B. V., Rusach, E. B., Setkina, V. N., Pyshnograeria, N. I.: J. Organometal. Chem. *77*, 69 (1974)	Raman (including crystal) and IR of η^5-$C_4H_4NMn(CO)_3$ and of η^5-$C_4D_4NMn(CO)_3$ Very similar to η^5-$C_5H_5Mn(CO)_3$
A79	Butler, I. S., Shaw, F. C.: J. Raman. Spec. *2*, 257 (1974)	The Raman spectra down to 15 K of $Mn^{12}(CO)_5Br$ and $Mn^{12}(CO)_4{}^{13}(CO)Br$. Not all of the expected isotopic bands appear

Table A (continued)

Ref.		
A80	Raissi-Shabari, A.: Q. Bull. Fac. Sci., Tehran Univ. *6,* 26 (1974)	The ν(CO) IR spectra of a variety of $Ni(CO)_{4-n}L_n$ and $Mo(CO)_{6-m}L_m$ complexes $n, m = 0, 1, 2, 3$ and L is one of the $P(ClC_6H_4)_3$ species
A81	Scorell, W. M., Spiro, T. G.: Inorg. Chem. *13,* 304 (1974)	Raman and IR spectra of solutions and crystals of $S_2Fe_2(CO)_6$ and $(CH_3S)_2Fe_2(CO)_6$ together with vibrational analyses
A82	Bor, G., Noack, K.: J. Organometal. Chem. *64,* 367 (1974)	The ν(CO) infrared spectrum of $Co_2(CO)_8$, enriched with ^{13}CO, indicates that three forms are present in solution
A83	Bor, G., Sbrignadello, G.: J. Chem. Soc. (Dalton) *1974,* 440	The IR spectra of $M_2(CO)_{10}$ M = Mn, Re, Tc with particular emphasis on ^{13}CO-containing species. A comparative vibrational analysis is given (including Raman data)
A84	Darensbourg, M. Y., Burns, D.: Inorg. Chem. *13,* 2970 (1974)	A ν(CO) IR study of solution of *trans* $Ph_3PFe(CO)_3C(O)Ph^-$ salts indicates ion pairs in Et_2O but not in T.H.F. or D.M.R.
A85	Lokshin, B. V., Aleksanya, V. T., Klemenkova, Z. S.: J. Organometal. Chem. *70,* 437 (1974)	The IR and Raman (solution and crystal) of η^2-(maleic anhydride)$Fe(CO)_4$
A86	Thornhill, D. J., Manning, A. R.: J. Chem. Soc. (Dalton) *1974,* 6	The ν(CO) IR spectra of a variety of $[L(Co(CO)_3]_2$ complexes, L = P, As or Ab ligand, show that solutions contain three isomers
A87	Hutchinson, B., Hance, R. L., Bernard, B. B., Hoffbauer, M.: J. Chem. Phys. *63,* 3694 (1975)	The IR spectrum of $^{54}Fe(CO)_5$ is compared with that of the natural-abundance compound
A88	Palyi, G., King, R. B.: Inorg. Chim. Acta *15,* L23 (1975)	Reduction in symmetry from C_{4v} of the $V(CO)_4$ unit in η^5-$C_5H_5V(CO)_4$ consequent upon acylation is observed in the ν(CO) IR spectrum
A89	Tripathi, S. C., Srivastava, S. C., Prasad, G., Mani, R. P.: J. Organometal. Chem. *86,* 229 (1975)	Vibrational data and their interpretation for several $AW(CO)_5$ and $A_2W(CO)_4$ species (A = amine)
A90	Palyi, G., Varadi, G.: J. Organometal. Chem. *86,* 119 (1975)	The IR spectra of $Co_3(CO)_9C$. A species (A = CH = CRCOOR′)
A91	Bee, M. W., Kettle, S. F. A., Powell, D. B.: Spect. Acta *31A,* 89 (1975)	The M-C-O vibrational features of $[M(NH_3)_5CO]Br_2$, M = Ru, Os are used in an analysis of the M-N_2 features of related compounds

Table A (continued)

Ref.		
A92	Braunstein, P., Dehand, J.: J. Organometal. Chem. *88*, C24 (1975)	Far IR data for M'-Au-M' systems, M' = $Mn(CO)_5$, $Co(CO)_4$, η^5-$C_5H_5Mo(CO)_3$ and η^5-$C_5H_5Fe(CO)_2$
A93	Lokshin, B. V., Rusach, E. B., Konovalov, Y. D.: Izv. Akad. Nauk SSSR *1975*, 84	Raman and infrared of solutions and crystalline η^6-$C_5H_5SCr(CO)_3$ are discussed and compared with η^6-$C_6H_6Cr(CO)_3$
A94	Onaka, S.: Nagoya Kogyo Daigaku Gakuho *25*, 133 (1973)	Raman study of metal-metal bonding spectral features of X_3M-$Mn(CO)_5$, M = Ge, Sn, X = Cl, CH_3
A95	Van den Berg, G. C., Oskam, A.: J. Organomet. Chem. *91*, 1 (1975)	Raman, IR and analyses of $X_3SiM'(CO)_4$, M' = Co, Fe, X = H, F, Cl
A96	Butler, I. S., Shaw, C. F.: J. Raman. Spect. *3*, 65 (1975)	Raman and IR solution and crystal spectra of $W(CO)_4NOBr$ are analyzed (data from natural abundance ^{13}CO-containing species are included)
A97	Doyle, G.: J. Organometal. Chem. *84*, 323 (1975)	The preparation, IR spectra and C.K. analysis of $M(CO)_5R$ species M = Cr, Mo, W; R = carboxylate
A98	Bor, G., Sbrignadello, G., Noack, K.: Helv. Chim. Acta *58*, 815 (1975)	The ν(CO) IR spectra of $Co_4(CO)_{12}$, $Rh_4(CO)_{12}$ and $HFeCo_3(CO)_{12}$ are analyzed using spectra of ^{13}CO-enriched compounds
A99	Harris, D. C., Gray, H. B.: J. Amer. Chem. Soc. *97*, 3073 (1975)	The IR spectra of $A[W(CO)_5]_2^+$, A = H, D are reported and discussed
A100	Swanson, B. I., Rafalko, J. J., Shriver, D. F., San Filippo, J., Spiro, T. G.: Inorg. Chem. *14*, 1737 (1975)	ν(metal-metal) in the Raman spectra of $Fe_2(CO)_9$, $Fe_3(CO)_{12}$, $Re_2(CO)_{10}$, $Re_2(CO)_8X_2$ (X = Br, Cl). Corrects or queries previous reports
A101	Adams, D. M., Christopher, R. E., Stevens, D. C.: Inorg. Chem. *14*, 1562 (1975)	The IR and Raman of η^6-$C_6A_6Cr(CO)_3$, A = H, D together with an analysis
A102	Caillet, P., Jaouen, G.: J. Organometal. Chem. *91*, C53 (1975)	Raman and IR of $ArCr(CO)_2CS$, Ar = methyl benzoate
A103	Johnson, J. R., Duggan, D. M., Risen, W. M.: Inorg. Chem. *14*, 1053 (1975)	Raman and IR of solutions and crystalline samples of $[ReM(CO)_{10}]^-$, M = Cr, Mo, W are analyzed and the results compared with those for analogous species
A104	Meester, M. A. M., Stufkens, D. J., Vrieze, K.: Inorg. Chim. Acta *14*, 25 (1975)	Raman and IR spectra of $PtCl_2(CO)L$ (L = a 4-pyridine ligand). ν(CO) shows a small dependence on the substituent in the pyridine ring

Table A (continued)

Ref.		
A105	Sbrignadello, G., Batiston, G., Bor, G.: Inorg. Chim. Acta *14,* 69 (1975)	The ν(CO) IR spectrum of $MnRe(CO)_{10}$ and its analysis, with emphasis on features derived from ^{13}CO-containing molecules
A105	Bigorgne, M.: J. Organometal. Chem. *94,* 161 (1975)	A review of the vibrational spectra of metal carbonyls. No references
A107	Hall, S. K.: Trans. III. State Acad. Sci. *67,* 250 (1974)	Synthesis and ν(CO) IR spectra of $R_{4-n}Sn[Co(CO)_3PPh_3]_n$ species, R = halide, alkyl or Ph
A108	Bigorgne, M., Kahn, O., Koenig, M. F., Lontellier, A.: Spect. Acta *31A,* 741 (1975)	Raman and IR Spectra of crystalline and glass solution of $C_6H_6Cr(CO)_3$ used, together with ^{13}CO data, in a vibrational analysis
A109	Chenskaya, T. B., Leites, L. A., Kriskaya, I. I., Babakhina, G. M.: Izv. Akad. Nauk. SSSR *1975,* 1292	The Raman and IR spectra of a variety of allyl $Fe(CO)_n$, n = 2,3-derivatives are reported and analyzed
A110	Bradley, G. F., Stobart, S. R.: Chem. Comm. *1975,* 325	Raman and IR at 90 K of $H_2M(CO)_4$, M = = Fe,Ru,Os
A111	Jeanne, C., Pince, R., Poilblanc, R.: Spect. Acta *31A,* 819 (1975)	Raman and IR spectra of $M(CO)_5NH_3$, M = Cr,Mo,W together with an analysis
A112	Rudnerskii, N. K., Vyshinskii, N. N., Grinval'd, I. I., Artemov, A. N., Sirotkin, N. I.: Tr. Khim. Khim. Tekhnol. *1974,* 127	A study of the effect on the IR spectra of substituents R in η^6-$C_6H_5RCr(CO)_3$
A113	Demuth, R., Grobe, J., Rau, R.: Z. Naturforsch. *30B,* 539 (1975)	The gas and liquid phase IR and Raman spectra of $(CF_3)_2AMn(CO)_5$ A = P,As are analyzed
A114	Butler, I. S., Spendjian, H. K.: J. Organometal. Chem. *101,* 92 (1975)	The ν(CO) IR spectra of $^{12}C^{18}O$ substituted *cis* $Mn(CO)_4LBr$ (L = PPh_3, $AsPh_3$, $SbPh_3$) and their analysis
A115	Sanger, A. R.: Chem. Comms. *1975,* 893	Infrared (and ^{31}P NMR) data show that $(RhCl(CO)\text{-}[Ph_2P(CH_2)_nPPh_2]_x$ are square planar with x = 2 except for n = 2 when x = 1
A116	Parker, D. J.: Spectrochim. Acta *31A*, 1789 (1975)	Raman and IR spectra of η^5-$MeC_5H_4Mn(CO)_3$ and $[\eta^5\text{-}MeC_5H_4Mo(CO)_3]_2$, including deuteration data, are assigned
A117	Arabi, M. S., Mathieu, R., Poilblanc, R.: J. Organometal. Chem. *104,* 323 (1976)	^1H NMR and IR spectra indicate stereochemical non-rigidity in $[HM(CO)_{6-n}(PR_3)_n]^+$ species M = Mo,W

Table A (continued)

Ref.		
A118	Ernstbrunner, E. E., Kilner, M.: J. Chem. Soc. (Dalton) *1975*, 2598	Raman and IR data for $Hg[Co(CO)_4]_2$ including intensity data, and their interpretation
A119	Onaka, S., Shriver, D. F.: Inorganic Chemistry *15*, 915 (1976)	Raman data on ν(M-M) in $Co_2(CO)_8$, $Fe_2(CO)_8^{2-}$, $Co_4(CO)_{12}$ and $[\eta^5\text{-}C_5H_5)Ru(CO)_2]_2$
A120	Barna, G. G., Butler, I. S., Plowman, K. R.: Can. J. Chem. *54*, 110 (1976)	Gas phase, solution and solid state infrared and Raman spectra of $\eta^5\text{-}C_5H_5Mn(CO)_{2-x}(CS)_{1+x}$, x = 0,1 and their assignment
A121	Wuyts, L. F., van der Kelen, G. P.: Spect. Acta *32A*, 689 (1976)	The far IR spectra of the species $R_nCl_{m-n}Sn[Co(CO)_4]_{4-m}$, R = alkyl, Ph, with particular emphasis on ν(Sn-Co)
A122	Butler, I. S., Shaw, C. F.: J. Mol. Str. *31*, 359 (1976)	Gas phase IR spectrum of $Cr(CO)_5Cs$ · PQR separations are seen on ν(CO) and the Coriolis coupling constant of the ν(CO) mode estimated
A123	Onaka, S., Shriver, D.: Inorg. Chem. *15*, 915 (1976)	Raman studies indicate that for first row transition metals ν(metal-metal) are above 200 cm^{-1} for CO bridged and between 190 and 140 cm^{-1} for unbridged compounds
A124	Barna, G. G., Britley, I. S., Plowman, K. R.: Can. J. Chem. *54*, 110 (1976)	The vibrational spectra of $\eta^5\text{-}C_5H_5Mn(CO)_2CS$ (vapour, solution and solid) and of $\eta^5\text{-}C_5H_5MnCO(CS)_2$ (solid)

Table B

Ref.		
B1	Buttery, H. J., Kettle, S. F. A., Keeling, G., Stamper, P. J., Paul, I.: J. Chem. Soc. (A) *1971*, 3148	Factor group splitting of ν(CO) in (dien)$M(CO)_3$ M = Cr, Mo, W, methyl substituted benzene $Cr(CO)_3$ and η^5-$C_5H_3Mn(CO)_3$
B2	Bullitt, J. G., Cotton, F. A.: Inorg. Chim. Acta *5*, 637 (1971)	A dipole-dipole model reproduces factor group frequencies of ν(CO) in a variety of dinuclear metal carbonyls
B3	Poletti, A., Paliani, G., Cataliotti, R., Toffani, A., Santucci, A.: J. Organometal. Chem. *43*, 377 (1972)	Single crystal IR study of $Mn(CO)_4NO$
B4	Cataliotti, R., Paliani, G., Poletti, A., Murgia, S. M., Cardaci, G.: Proc. Convegno. Naz. Chim. Inorg. *1973*, 369	The polarized infrared spectra of crystalline η^3-$C_3H_5Fe(CO)_2(NO)$ and η^3-$C_3H_5Co(CO)_3$
B5	Adams, D. M., Fernando, W. S., Hooper, M. A.: J. Chem. Soc. (Dalton) *1973*, 2264	Solution and single-crystal Raman spectra of $M(CO)_6$, M = Cr, Mo, W
B6	Lokshin, B. V., Aleksanyan, V. T., Klemenkova, Z. S.: J. Organometal. Chem. *70*, 437 (1974)	Raman and IR of solution and crystalline samples of (η^2-maleic anhydride) $Fe(CO)_4$ and their assignment
B7	Adams, D. N., Trumble, W. R.: J. Chem. Soc. (Dalton) *1974*, 690	The IR spectra of single crystals of $Rh(CO)_2L$ L = acetylacetonato
B8	Buttery, H. J., Kettle, S. F. A., Paul, I.: J. Chem. Soc. (Dalton) *1974*, 2293	Single crystal Raman study of the $Cr(CO)_3$ group in $RCr(CO)_3$, R = C_6H_6, 1,3,MeC_6H_4
B9	Kahn, O., Bigorgne, M., Koenig, M. F., Loutellier, A.: Spect. Acta *30A*, 1929 (1974)	Raman and IR spectra of crystalline $Fe(CO)_4AMe_3$ (A = P, As, Sb), $Co(CO)_4L'$ (L = Si, $GeCl_3$, Sn, Pb, Me_3). A vibrational unit cell of twice the size of that of the crystallographic was deduced for some species
B10	Buttery, H. J., Kettle, S. F. A., Paul, I.: J. Chem. Soc. (Dalton) *1975*, 969	Single crystal Raman spectra of $ArCr(CO)_3$ Ar = C_6Me_6, HC_6Me_5

Table C

Ref.		
C1	Fenske, R. F.: Pure Appl. Chem. *27,* 61 (1971)	The interpretation of ν(CO) force constants by M. O. theory
C2	Brunvoll, J., Cyvin, S. J., Schaefer, L.: J. Organometal. Chem. *36,* 143 (1972)	Normal coordinate analysis of η^6-$C_6H_6Cr(CO)_3$ and mean amplitude etc.
C3	Kjelstrup, S., Cyvin, S. J., Brunvoll, J., Schaefer, L.: J. Organometal. Chem. *36,* 137 (1972)	Various normal coordinate analyses of η^6-$C_6H_6Cr(CO)_3$
C4	Derarajan, V., Cyvin, S. J.: Act. Chem. Scand. *26,* 1 (1972)	Normal coordinate analysis and mean amplitudes etc., of $Cl_3MCo(CO)_4$, M = Si, Ge, Sn
C5	Hall, M. B., Fenske, R. F.: Inorg. Chem. *11,* 1619 (1972)	M. O. calculations of the occupancy of the relevant CO σ and π orbitals in a variety of complexes correlated with force constant data
C6	Dennenberg, R. J., Darensbourg, D. J.: Inorg. Chem. *11,* 72 (1972)	Neither the rate constant nor amine pK_α in the thermal decomposition of $M(CO)_5A$, M = Cr,M,W, A = amine is correlated with ν(CO) force constants
C7	Brown, R. A., Dobson, G. R.: Inorg. Chim. Acta *6,* 65 (1972)	Correlation of M-C-O-derived frequencies with ligand in $LM(CO)_5$ M = Cr,Mo,W, L = Lewis base
C8	Angelici, R. J., Blacik, L. J.: Inorg. Chem. *11,* 1754 (1972)	Proposed relationship between ν(CO) force constants and reactivity
C9	Brill, T. B.: J. Organometal. Chem. *40,* 373 (1972)	Proposed correlation between ν(CO) IR spectra and Mn electron density in η^5-$C_5H_5Mn(CO)_2$-$X(OR_3)$, X = As,Sb
C10	Jones, L. H., McDowell, R. S., Goldblatt, M., Swanson, B. I.: J. Chem. Phys. *57,* 2050 (1972)	A detailed vibrational analysis of $Fe(CO)_5$ using data from fully $^{13}C^{16}O$, $^{12}C^{18}O$ and $^{12}C^{16}O$ substituted species
C11	Chen, C., Hsiang, C. C.: J. Chin. Chem. Soc. *20,* 13 (1973)	A suggested ν(CO) vibrational analysis technique using a maximum interaction model to overcome insufficiency of data
C12	Van den Berg, G. C., Oskam, A.: J. Organometal. Chem. *78,* 357	Vibrational analyses of species in the series $X_3MCo(CO)_4$, M = Si,Ge,Si. X = H,D,F,Cl, Br,I
C13	Bor, G.: J. Organometal. Chem. *65,* 81 (1974)	The interpretation of the infrared spectra of $(CO)_5M \cdot Co(CO)_4$ M = Mn,Tc,Re requires a model with free rotation around the M-Co bond

Table C (continued)

Ref.		
C14	Bodner, G. M., Todd, L. J.: Inorg. Chem. *13*, 1335 (1974)	A correlation between ^{13}C NMR shifts and $\nu(CO)$ in a variety of carbonyls [especially of η^6-arene $Cr(CO)_3$]
C15	Dobson, G. R.: Ann. N. Y. Acad. Sci. *1974*, 239	From a comparison of kinetic and spectral data it is concluded that in CO substitution reactions of $Mn(CO)_5X$, the positions of ν(M-CO) are diagnostic of reactivity
C16	Emsley, J. W., Lindon, J. C.: Mol. Phys. *28*, 1373 (1974)	A vibrational analysis (excluding vibrations of the MCO unit) is used in interpreting the results of a 'H NMR, nematic solution study of η^4-$C_4H_4Fe(CO)_3$
C17	Bor, G.: J. Organometal. Chem. *94*, 181 (1975)	A review of the effect bridging ligands on k(CO) in $Co_2(CO)_6L_2$, $Fe_2(CO)_6L_2$, (L = bridging ligand)
C18	Bigorgne, M.: Spect. Acta *31A*, 317 (1975)	The ν(CO) Raman and IR spectra of $Mo(CO)_4(PEt_3)_2$, $Fe(CO)_3[P/OMe)_3]_2$ and $Ni(CO)_4(CO)_4$ in solution, solid and glass are used to calculate force constants. The solid-state data play an important part in the analysis. Uses isotopic data
C19	Andronov, E. A., Kukuskin, Y. N., Churakov, V. G., Murashkin, Y. V.: Zh. Neorg. Khim. *20*, 1126 (1975)	A correlation is reported between ν(CO) and ν(Pd-C) in the series $PdS(CO)Cl_2$, S = sulphur containing ligand
C20	Andrews, D. C., Davison, G., Duce, D. A., J. Organometal. Chem. *97*, 95 (1975)	An analysis of the vibrational spectra of $C(CH_2)_3Fe(CO)_3$
C21	Neuse, E. W.: J. Organometal. Chem. *99*, 287 (1975)	Correlation between σ parameters and k(CO) for substituted arene $Cr(CO)_3$ complexes
C22	Edgell, W. F.: Spectrochim. Acta *31A*, 1623 (1975)	A proposal that a hidden-dual-symmetry is responsible for the failure of $Mn(CO)_5Br$ to obey the C_{4v} selection rules in the ν(CO) region
C23	Brownlee, R. T. C., Topson, K. D.: Spect. Acta *31A*, 1677 (1975)	Includes an analysis of substituent-parameter correlations for η^5-$C_5H_5Fe(CO)_2X$ for a variety of X
C24	Lokshin, B. V., Rusach, E. B., Kaganovich, V. S., Krivykh, V. V., Artemov, A. N., Sirotkin, N. I.: Zh. Strukt. Khim. *16*, 592 (1975)	Raman and IR of $RM(CO)_3$ M = Cr, Mo, W, R = C_6H_6, 1,3,5-$Me_3C_6H_3$, are analyzed and correlated with a σ parameter

Table C (continued)

Ref.		
C25	Caillet, P.: C. R. Hebd. Seances Acad. Sci. *281*, 1057 (1975)	Discussion of the force field of $RCr(CO)_2(CS)$ R = methyl benzoate
C26	Barnett, G. H., Cooper, M. K.: Inorg. Chim. Acta *14*, 223 (1975)	A method for determining anharmonicity corrections to C. K. force field is proposed
C27	Jones, L. H., Swanson, B. I.: Accounts of Chemical Res. *9*, 128 (1976)	The interpretation of potential constants obtained from vibrational analyses of metal carbonyls (and cyanides)
C28	Pince, R., Poilblanc, R.: J. Chim. Phys. *72*, 1087 (1975)	Normal coordinate analyses of $[RhCl(CO)_2]_2$, $[RhBr(CO)_2]_2$ and $[RhCl(CO)PMe_3]_2$
C29	Caillet, P.: J. Organometall. Chem. *102*, 481 (1975)	Relationship between the force fields of methyl benzoate and $Cr(CO)_6$ and that of η^6-methyl benzoate $Cr(CO)_3$

Table D

Ref.		
D1	Keeling, G., Kettle, S. F. A., Paul, I.: J. Chem. Soc. (*A*), 3143 (1971)	Absolute IR intensities of ν(CO) in $XM(CO)_5$ M = Mn, Re and $XM(CO)_5^-$, M = Cr, W, X = Cl, Br, I
D2	Anderson, W. P., Brown, T. L.: J. Organometal. Chem. *32*, 343 (1971)	Simple M.O. calculations on $M(CO)_5L$ and fac.-$M(CO)_3L_3$ indicate that ν(CO) intensities are a measure of π bonding
D3	Darensbourg, D. J.: Inorg. Chem. *11*, 1606 (1972)	Integrated ν(CO) infrared intensities and bond angle in the $Fe(CO)_2$
D4	Schlodder, R., Vogler, S., Beck, W.: Z. Natur. *27B*, 27 (1972)	Integrated ν(CO) intensities in monocarbonyls of Ir and Rh; correlation with a scale of ligand π-donor ability
D5	Kettle, S. F. A., Paul, I.: Adv. Organometal. Chem. *10*, 199 (1972)	Review of ν(CO) infrared intensities
D6	Koenig, M. F., Bigorgne, M.: Spect. Acta *28A*, 1693 (1972)	Measurement and interpretation of Raman intensities of $Ni(CO)_{4-x}L_x$ species, L = P ligand
D7	Anderson, W. P., Brill, T. B., Schoenberg, A. R., Stanger, Jr., C. W.: J. Organometal. Chem. *44*, 161 (1972)	Infrared intensities of a variety of derivatives of η^5-$C_5H_5Mn(CO)_3$ correlated with a simple MO model. Also ^{55}Mn Mössbauer data
D8	Kettle, S. F. A., Paul, I., Stamper, P. J., J. Chem. Soc. (Dalton) *1972*, 2413	Interpretation of the ν(CO) Raman intensities of a variety of carbonyls using a bond polarizability model
D9	Koenig, M. F., Bigorgne, M.: Adv. Raman Spect. *1*, 563 (1972)	Raman intensities of $Ni(CO)_{4-n}L_n$, L = = phosphorus ligand, and their interpretation
D10	Hyde, C. L., Darensbourg, D. J.: Inorg. Chem. *12*, 1075 (1973)	Absolute ν(CO) IR band intensity measurements and interpretation for a variety of *cis* $W(CO)_4LL'$ species
D11	Samrelyan, S. K., Aleksanyova, V. T., Lokshin, B. V.: J. Mol. Spec. *48*, 47 (1973)	Absolute infrared intensities of $M(CO)_6$, M = Cr, Mo, W (all infrared active modes are studied)
D12	Onaka, S.: J. Inorg. Nucl. Chem. *36*, 1721 (1974)	The infrared intensities of the ν(CO) bands in X_3M-$Mn(CO)_5$, M = Si, Ge, Sn and X = Cl, Br, I
D13	Schoenberg, A. R., Anderson, W. P.: Inorg. Chem. *13*, 465 (1974)	ν(CO) infrared intensity measurements for a variety of $Mn(CO)_2$-containing species (also NMR studies)

Table D (continued)

Ref.		
D14	Onaka, S.: J. Inorg. Nucl. Chem. *36*, 1721 (1974)	Absolute intensities of the ν(CO) IR bands in $X_3MMn(CO)_5$, M = Si, Ge, Sn; X = Cl, Br, I, and their correlation with force constants
D15	Butler, I. S., Johansson, D. A.: Inorg. Chem. *14*, 701 (1975)	Absolute integrated intensity data for ν(CO) and ν(CS) in η^5-$C_5H_5Mn(CO)_2CS$
D16	Sizova, O. V., Ivanova, N. V., Baranovskii, V. I., Nikol'skii, A. B.: Opt. Spekt. *39*, 1086 (1975)	Application of a semi-empirical MO method to the prediction of infrared intensities of AB = CO, N_2, CN and NO in $[Os(NH_3)_3Cl_2AB]^{n+}$
D17	Paetzold, R., A-El-Mottalab: J. Mol. Str. *24*, 357 (1975)	The absolute IR intensities of ν(CO) features, bond polarizations (MC and CO) and other quantities (^{13}C chemical shifts, charge-transfer spectra, calculated CO 2pπ populations) are correlated
D18	Ahmed, M. S., A-El-Mottalab: J. Mol. Struct. *25*, 438 (1975)	Absolute IR intensities for $M(CO)_{6-x}X_x$, M = Cr, Mn, Fe, X = Cl, Br, I, X = 1,2 are reported and discussed as reflecting bonding properties
D19	Bigorgne, M.: Spect. Acta *31A*, 1151 (1975)	A detailed analysis of bond moment derivatives of $Ni(CO)_3PMe_3$ from absolute intensity data
D20	Abdul-El-Mottaleb: J. Mol. Str. *32*, 203 (1976)	Correlation between ν(CO) infrared intensities and ^{183}W- ^{31}P spin-spin coupling constants for a variety of $R_3PW(CO)_5$ complexes

Table E

Ref.		
E1	Knox, A. R. S., Hoxmeier, R. J., Kaesz, H. D.: Inorg. Chem. *10*, 2636 (1971)	Dangers of CCl_4 as a solvent. Reaction with η^5-$C_5H_5(CO)_3W$-$Mn(CO)_5$ and $MnRe(CO)_{10}$
E2	Canini, S., Ratcliff, B., Fusi, A., Pasini, A.: Gann. Chim. Ital. *102*, 141 (1972)	Solvent effects on ν(CO) in a variety of carbonyls with metal-metal bonds
E3	Spaulding, L., Reinhardt, B. A., Orchin, M.: Inorg. Chem. *11*, 2092 (1972)	Reaction of a Pt(II) carbonyl derivative with halide from the IR cell window monitored by ν(CO)
E4	Wiggans, P. W.: Educ. Chem. *10*, 52 (1973)	Infrared ν(CO) studies suitable for undergraduate teaching
E5	Ludlum, K. H., Eischens, R. P.: Surface Sci. *40*, 397 (1973)	Gaseous CO attacks a stainless steel cell to give metal carbonyl peaks within five minutes
E6	Knebel, W. J., Angelici, R. J., Gansow, O. A., Darensbourg, J.: J. Organometal. Chem. *66*, C11 (1974)	Stereospecific replacement of ^{12}CO by ^{13}CO in $LMo(CO)_4$ complexes (L = P-N or P-P bidentate ligand) followed by IR (some spectral data given)
E7	Pribula, C. D., Brown, T. L.: J. Organometal. Chem. *71*, 415 (1974)	Infrared evidence for ion pairs in $MMn(CO)_5$, M = Li, Na in ether and THR. Also effect of added Mg^{2+} and $LiCo(CO)_4$
E8	Mahnke, H., Clark, R. J., Rosanske, R., Sheline, R. K.: J. Chem. Phys. *60*, 2997 (1974)	An IR study of the relative proportions of the various $Fe(CO)_{5-x}(PF_3)_x$ species as a f (temperature) used to provide data for a corresponding ^{13}C NMR study
E9	Hsieh, A. T. T., West, B. O.: J. Organometal. Chem. *78*, C40 (1974)	A solvent dependence of ν(CO) infrared frequencies is observed for some allyl-$Mo(CO)_2XL_2$ and -$W(CO)_2XL_2$ species L = N-ligand (electronic spectra also show a dependence)
E10	Gould, N. J., Parker, D. J.: Spectrochim. Acta *31A*, 1785 (1975)	The ν(CO) solvent dependence in $Mn_2(CO)_{10}$ $Mn_2(CO)_8L_2$, L = P ligand is, in non-polar solvents, linearly related to a solvent parameter
E11	Kopsch, A., Hellner, E., Dehnicke, K.: Bunsenges. Phys. Chem. Ber. *80*, 500 (1976)	The IR spectra of a variety of metal carbonyl derivatives (and $[Fe(CN)_5NO]^{2-}$ and N_3^- salts) under high pressure. Little change in CO symmetry is generally observed

Inorganic Applications of X-Ray Photoelectron Spectroscopy

Prof. William L. Jolly

Department of Chemistry, University of California, and the Materials and Molecular Research Division, Lawrence Berkeley Laboratory, Berkeley, California 94720, U. S. A.

Table of Contents

I. Introduction

The principles of X-ray photoelectron spectroscopy (XPS or ESCA) have been covered in books by Siegbahn *et al.*[1, 2] and Carlson[3]. Comprehensive reviews of the inorganic aspects of the field appeared in the 1972, 1973, and 1974 volumes of *Electronic Structure and Magnetism of Inorganic Compounds*[4] in the Chemical Society series of Specialist Periodical Reports and will presumably appear in future volumes of this publication. Reviews covering applications of XPS to coordination chemistry[5] and the equivalent cores approximation[6] were recently published. The XPS literature is classified and briefly summarized in biennial reviews by Hercules *et al.*[7]. The purpose of this chapter is to describe recent XPS research involving problems of interest to inorganic chemists. Most of the chapter is devoted to methods for systematizing binding energy data and for deducing information about chemical bonding from the spectra. Some poorly understood phenomena and perplexing aspects of the technique are discussed to point out promising areas for future work.

II. Methods for Predicting Chemical Shifts

Chemists are interested in X-ray photoelectron spectroscopy mainly because the measured electron binding energy of an atomic core level is a function of the chemical environment of the atom. It has been shown that chemical shifts in the core binding energy of an atom in various compounds can be correlated with properties such as the nature of the groups bonded to the atom, atomic charges, thermodynamic data, and the electronic structures of the compounds. A correlation of binding energy with one of these properties can be of practical value in the case of a compound for which the property is unknown or uncertain. In that case an appropriate experimental binding energy of the compound can be used, with the aid of the established correlation, to estimate that property for the compound. In this section we shall discuss and evaluate the methods for predicting chemical shifts in binding energy which are of value to chemists.

A. Empirical Group Parameters

The very simple idea that chemical shifts in binding energy can be accounted for in terms of changes in substituent groups is the basis of a predictive method involving additive group shift parameters. The method may be represented by the equation

$$\Delta E_B = \sum_i \Delta E_{gr}(i) \qquad (1)$$

where ΔE_B is the binding energy shift and $\Delta E_{gr}(i)$ is the part of the shift due to the group i attached to the atom studied. A simple example will make the method clear.

The absolute and relative (to CH_4) carbon 1s binding energies for CH_4, CF_4, CCl_4, and $CClF_3$ are given in Table 1. From the data for CF_4 and CCl_4 we can calculate group shift parameters of 2.76 and 1.38 eV for fluorine and chlorine atoms, respectively.

Table 1. Carbon 1s binding energies

Compound	E_B, eV	ΔE_B, eV	Ref.
CH_4	290.71	0.00	8)
CF_4	301.76	11.05	8)
CCl_4	296.22	5.51	8)
$CClF_3$	300.13	9.42	8, 9)

Using these parameters, we calculate $\Delta E_B = 1.38 + 3 \times 2.76 = 9.66$ eV for $CClF_3$, in fair agreement with the experimental value. Obviously the method assumes that the binding energy of a chlorofluorocarbon is the weighted average of the binding energies of CF_4 and CCl_4. Separate sets of group shift parameters have been calculated by least squares treatments of a large number of carbon 1s[10, 11], nitrogen 1s[12], silicon 2p[13], phosphorus 2p[12, 14–16], and arsenic 3p[12] binding energy data. In general, the experimental values can be reproduced by Eq. (1) with an average deviation of about 0.3 eV, which is approximately the experimental uncertainty in the binding energies. However, these results are not very impressive when one considers the large number of empirical parameters used. For example, in the correlation of 104 different nitrogen 1s binding energies[12], 57 group shift parameters were required. Of these parameters, 22 were determined from single binding energies and therefore automatically yielded perfect predictions.

Lindberg *et al.*[17] have shown that the carbon 1s shifts of some substituted benzenes are linearly correlated with the Hammett σ parameters of the substituents. However, Hammett parameters are of rather limited applicability; a given set of σ values can be used to correlate data only for similar chemical systems. It has recently been shown[18] that a wider variety of core binding energy shifts can be correlated by the four-parameter relation

$$\Delta E_B = aF + bR \qquad (2)$$

in which the parameters a and b are characteristic of the class of molecule and atom to which the binding energies pertain and the parameters F and R are characteristic of substituent groups. For example, there are particular values of a and b which correspond to the carbon 1s binding energies of compounds of the type CH_3X, relative to methane. Similarly, there are particular values of F and R for the chloro group. Substitution of these values in Eq. (2) yields the relative binding energy for methyl chloride. Sixteen pairs of a and b values have been evaluated corresponding to the core ionizations $\overset{*}{C}H_3X$, $\overset{*}{C}F_3X$, $O\overset{*}{C}X_2$, $\overset{*}{C}X_4$, $\overset{*}{C}H_2CHX$, $\overset{*}{F}X$, $\overset{*}{F}_3CX$, $\overset{*}{B}X_3$, $\overset{*}{Si}X_4$, $\overset{*}{Ge}X_4$, $\overset{*}{Sn}X_4$, $\overset{*}{P}X_3$, $\overset{*}{O}CX_2$, $\overset{*}{Cl}X$, $\overset{*}{Br}X$, and $\overset{*}{I}X$, and ten pairs of F and R values have been evaluated corresponding to the substituents CH_3, CF_3, C_6H_5, SiH_3, GeH_3, OCH_3, F, Cl, Br, and I, in addition to H, the reference substituent.

Forty-eight of these parameters are adjustable and were evaluated by a least-squares analysis of 92 experimental binding energies. The average deviation between the experimental and calculated ΔE_B values was ±0.2 eV.

The *F* and *R* parameters are qualitatively analogous to the "field" and "resonance" parameters, $\mathfrak{F}$ and $\mathfrak{R}$, of Swain and Lupton[19]; that is, they measure the σ and π-electronegativities, respectively, of substituents. However, the *F* and *R* values are more appropriate for correlating processes in which a localized positive charge develops than are the $\mathfrak{F}$ and $\mathfrak{R}$ values. Hence the *F* and *R* values correlate lone pair ionization potentials and proton affinities better than the corresponding $\mathfrak{F}$ and $\mathfrak{R}$ values do.

In principle, the empirical group parameter methods can be used to aid molecular structure determination, but their principal use to date has been in the determination of the nature of bonding and the electron distribution in compounds by interpretation of the magnitudes of the empirically evaluated parameters.

B. The Equivalent Cores Approximation

A core electron binding energy E_B can be considered as the energy of a chemical reaction. In the case of the carbon 1s binding energy of gaseous methane, the reaction is

$$CH_4 \longrightarrow CH_4^{+*} + e^- \tag{3}$$

in which the asterisk indicates the absence of one of the carbon atom's 1s electrons. The heats of formation of CH_4 and e^- are well known, but we cannot directly obtain the heat of formation of CH_4^{+*} from ordinary thermodynamic data. However, we can do this if we apply the so-called "equivalent cores" approximation[6, 11, 20–24]. According to this approximation, the hypothetical process in which an electron is transferred from the nucleus of a core-ionized atom to the core hole has an energy which is independent of the chemical environment of the core-ionized atom. For example, in the case of carbon 1s holes in CH_4 and CO_2, it is assumed that the following reactions have the same energy.

$$CH_4^{+*} \longrightarrow NH_4^+ \tag{3a}$$

$$CO_2^{+*} \longrightarrow NO_2^+ \tag{3b}$$

(In these reactions it is understood that the C and N atoms have the same mass numbers.) The approximation that reactions of this type have the same energy is equally valid if we convert the reactions into ordinary chemical reactions by adding $1/2\,N_2$ and $C_{(graphite)}$ to the left and right sides, respectively, of each reaction:

$$CH_4^{+*} + 1/2\,N_2 \longrightarrow NH_4^+ + C_{(graphite)} \tag{4a}$$

$$CO_2^{+*} + 1/2\,N_2 \longrightarrow NO_2^+ + C_{(graphite)} \tag{4b}$$

For such core replacement reactions the ordinary isotopic distributions of the elements can be assumed. It is assumed that the energy of any reaction like (4a) or (4b) is a constant, Δ_C. Note that if we add reactions as follows, (3a) + (4a) and (3b) + (4b), we eliminate the core-ionized species:

$$CH_4 + 1/2\,N_2 \longrightarrow NH_4^+ + C_{(graphite)} + e^- \tag{5a}$$

$$CO_2 + 1/2\,N_2 \longrightarrow NO_2^+ + C_{(graphite)} + e^- \tag{5b}$$

The energies of reactions (5a) and (5b) are $E_B(CH_4) + \Delta_C$ and $E_B(CO_2) + \Delta_C$, respectively. Hence the constant Δ_C may be calculated from the reactions

$$\Delta_C = \Delta H_f^\circ(NH_4^+) - \Delta H_f^\circ(CH_4) - E_B(CH_4) \tag{6a}$$

$$\Delta_C = \Delta H_f^\circ(NO_2^+) - \Delta H_f^\circ(CO_2) - E_B(CO_2) \tag{6b}$$

In the general case of a molecule $M(Z)$ containing an atom of atomic number Z which undergoes core ionization,

$$\Delta_Z = \Delta H_f^\circ[M(Z+1)^+] - \Delta H_f^\circ[M(Z)] - E_B[M(Z)] \tag{7}$$

Obviously for each type of core ionization (C 1s, O 1s, P 2p, etc.), there is a different value of Δ_Z. Carbon 1s binding energies and the corresponding calculated Δ_C values for some gaseous carbon compounds are given in Table 2. It can be seen that the various Δ_C values are fairly constant;

Table 2. Carbon 1 s binding energies and corresponding values of the energies of the core replacement reactions, Δ_C

Cpd.	E_B, eV	Ref.	Δ_C, eV ($\Delta H_f^\circ[M(N)^+] - \Delta H_f^\circ[M(C)] - E_B$)
CHF_3	299.1	25)	−283.9
CO_2	297.71	26)	−283.69
HCN	293.5	27)	−283.9
C_2H_2	291.14	28)	−283.86
CH_4	290.88	8)	−283.49
C_2H_6	290.74	8)	−283.39
C_6H_6	290.42	17)	−283.46
$CH_3\overset{*}{C}CH$	290.40	28)	−283.71

the average value is −283.68 eV, and the average deviation from this value is ± 0.17 eV. Hence we may write, for carbon 1 s binding energies,

$$E_B(C\ 1s) = \Delta H_f^\circ(MN^+) - \Delta H_f^\circ(MC) + 283.68$$

Similar treatment of binding energies for gaseous compounds of boron, nitrogen, oxygen, phosphorus, and sulfur leads to the following relations.

$$E_B(\mathrm{B\ 1s}) = \Delta H_f^\circ(\mathrm{MC^+}) - \Delta H_f^\circ(\mathrm{MN}) + 186.9$$
$$E_B(\mathrm{N\ 1s}) = \Delta H_f^\circ(\mathrm{MO^+}) - \Delta H_f^\circ(\mathrm{MN}) + 399.37$$
$$E_B(\mathrm{O\ 1s}) = \Delta H_f^\circ(\mathrm{MF^+}) - \Delta H_f^\circ(\mathrm{MO}) + 529.62$$
$$E_B(\mathrm{P\ 2p_{3/2}}) = \Delta H_f^\circ(\mathrm{MS^+}) - \Delta H_f^\circ(\mathrm{MP}) + 127.7$$
$$E_B(\mathrm{S\ 2p}) = \Delta H_f^\circ(\mathrm{MCl^+}) - \Delta H_f^\circ(\mathrm{MS}) + 160.0$$

Obviously any one of these relations may be used to predict a core binding energy when the necessary heats of formation are known or may be used to predict an unknown heat of formation when the necessary binding energy and one heat of formation are known.

Application of the equivalent cores method to solid compounds is slightly more complicated, requires additional assumptions, and is therefore less accurate than the application to gaseous compounds. However, fairly good correlations have been obtained for solid compounds of boron, carbon, nitrogen, and iodine[20]. The correlations were restricted, because of the nature of the assumptions involved, to molecular compounds or to compounds in which the core-ionized atoms are in anions.

A chemical shift in core binding energy corresponds, according to the equivalent cores approximation, to the energy of an ordinary chemical reaction. For example, the difference between the carbon 1s binding energies of CH_2Cl_2 and CH_4 is equated to the energy of the following reaction.

$$CH_2Cl_2 + NH_4^+ \longrightarrow NH_2Cl_2^+ + CH_4$$

Because the heat of formation of $NH_2Cl_2^+$ is not known, this chemical shift cannot be calculated from available thermodynamic data. However, quantum mechanical methods have been used to estimate the energy of this and similar reactions. These calculations are attractive for at least three reasons:

1. Generally only closed-shell calculations are required,
2. the *difference* in energy between two isoelectronic species can be calculated more accurately than the *absolute* energy of either species, and
3. the calculations can be made for isoelectronic species having the same geometry, in accord with the fact that nuclei remain fixed during the photoelectric process.

In Table 3 experimental carbon 1s binding energy shifts (relative to CH_4) for several compounds can be compared with the energies of the corresponding "equivalent-cores" chemical reactions calculated by various quantum mechanical methods. The agreement with experiment improves, as expected, with increasing sophistication of the calculational method. The poorest agreement is obtained with the CNDO/2 method, and the best agreement is obtained with *ab initio* methods using Slater-type orbitals fitted by Gaussian functions. Although chemical shifts calculated by the semiempirical MO methods are in poor absolute agreement with the experimental values, the calculated values show fairly good linear correlations with the experimental values, particularly for compounds having similar structures[29].

Table 3. Some carbon 1s binding energy shifts calculated using the equivalent cores approximation

Cpd.	ΔE_B, eV					
	Exptl.	Thermo.	CNDO/2[29]	MINDO/1[30]	MINDO/3[31]	*Ab initio*
CH_4	0	0	0	0	0	0
CHF_3	8.2	7.8	1.68	–	10.78	9.31[32]
CF_4	11.05	11.3	2.00	–	14.48	12.64[32]
CO_2	6.83	6.6	–	2.33	4.88	5.6 [33]
CH_3OH	1.6	–	–	0.56	2.09	–
CH_2Cl_2	3.1	–	1.62	–	–	3.39[32]

C. Correlation with Atomic Charge

Simple electrostatic considerations lead to the conclusion that the energy required to remove a core electron from an atom should increase with increasing positive net atomic charge and decrease with increasing negative net atomic charge. This effect can be seen in the data of Table 4, where chemical shifts in the 1s binding energy of sulfur atoms with various charges (calculated by an SCF Hartree-Fock method[1]) are listed. By interpolation of such data it is possible to calculate binding energy shifts for hypothetical fractionally-charged atoms. Over a small range of atomic charge, binding energy is well represented as a linear function of atomic charge, as in the following equation,

Table 4. Calculated sulfur 1s shifts

Species	ΔE_B, eV
S^-	−11.3
S	0
S^+	13.8
S^{2+}	29.8
S^{3+}	48.1

$$E_B = kQ + l \qquad (8)$$

where Q is the atomic charge and k and l are constants. This same equation can also be derived by consideration of a simple shell model of the atom[1].

If one assumes that the core binding energy of an atom in a compound is dependent only on the atomic charge, one should expect Eq. (8) to serve as a basis for correlating binding energies in compounds. Indeed, many investigators have obtained fairly good correlations using Eq. (8) in combination with methods for estimating atomic charges. The method which has been most commonly used for estimating atomic charges is the Pauling[34] method (or modifications of it), which involves the assumption that the partial ionic character of a bond is given by the relation

$$I = 1 - e^{-0.25(\Delta x)^2}$$

where Δx is the difference between the electronegativities of the bonded atoms. For example, in this way Grim *et al.*[35, 36] found linear correlations between the calculated metal atom charges and the nickel $2p_{3/2}$ binding energies of some simple nickel (II) salts and the molybdenum $3d_{5/2}$ binding energies of compounds of molybdenum in a variety of oxidation states. Similarly Hughes and Baldwin[37] found a linear correlation between molybdenum atom charges in various triphenylphosphine molybdenum complexes and the molybdenum $3d_{3/2}$ binding energies.

The potential energy of a core electron is not only affected by the charge of the atom of which it is a part, but is also affected by the charges of all the other atoms in the compound. That is, one must account for the work to remove the electron from the field of the surrounding charged atoms as well as from the atom which loses the electron. Both effects are accounted for by the so-called point-charge potential equation[2]

$$E_B(A) = kQ_A + \sum_{i \neq A} (Q_i/R_i) + l \qquad (9)$$

in which Q_i are the charges of the atoms and R_i are their distances from the core-ionizing atom (atom A). When the more approximate equation [Eq. (8)] is used to correlate binding energies with atomic charge, good straight-line relationships are found only if the compounds involved have similar structures so that the omitted term,

$$\sum_{i \neq A} (Q_i/R_i)$$

is either constant or approximately linearly related to Q_A. In general, very little improvement in correlations with Eq. (8) is achieved by the use of more sophisticated methods for estimating atomic charges. However, when Eq. (9) is used, it is important to use a good charge estimation procedure, and the resulting correlations are usually superior to the corresponding correlations with Eq. (8). Gray *et al.*[13] have recently compared the correlations obtained for a set of silicon 2p binding energies using Eqs. (8) and (9) with several different methods for estimating atomic charges.

In salts, rather small core binding energy shifts are observed for monatomic ions and for atoms in polyatomic ions when the counter-ions are changed. For example, in a series of sixteen potassium salts, the spread in K 2p binding energy (between the extremes of KCl and $K_2[Pt(NO_2)_4Cl_2]$) is only 1.7 eV.[38]. And both the N 1s and P 2p binding energies of many salts containing the bis(triphenylphosphine) iminium cation, $N[P(C_6H_5)_3]_2^+$, differ by only a few tenths of an eV[39]. The minor effect of crystal environment on chemical shift, illustrated by these results, is surprising when one considers the magnitude of the potential term of Eq. (9)

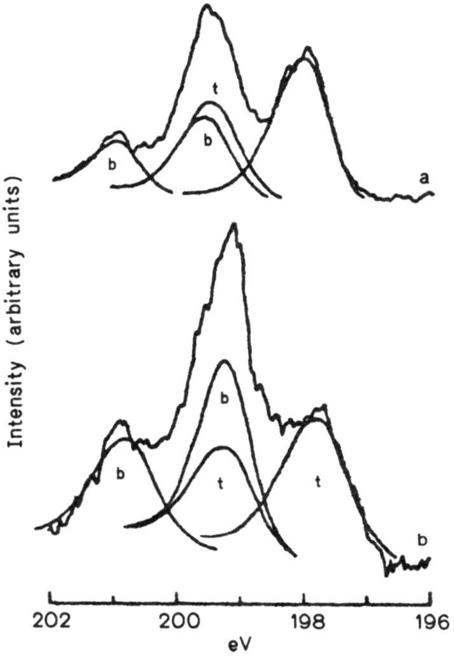

Fig. 1. Chlorine 2p spectra of (a) $Re_3Cl_9(pyz)_3$ and (b) $[Re_3Cl_6(py)_3]_n$ showing deconvolutions into two Cl $2p_{1/2, 3/2}$ doublets; bridging and terminal chlorine components are distinguished by the labels b and t. Reproduced with permission from Ref.[42]

as calculated using an appropriate Madelung function. In the case of simple salts such as KCl and LiF, if it is assumed that the ions bear unit charges, the chemical shifts calculated from the differences in the Madelung potentials differ greatly from the small observed chemical shifts. Citrin *et al.*[40] showed that the discrepancies can be markedly reduced by accounting for the mutual polarization of the ions. The inclusion of polarization effects is essentially equivalent to the use of covalent bonding with fractional atomic charges. In fact, Parry[41], by the solution of simultaneous equations [of the form of Eq. (9)] has calculated atomic charges in several oxides of lead using theoretical k values.

Hamer and Walton[42] have shown that it is possible to distinguish, within a single compound, chlorine atoms which are bonded to only one metal atom and chlorine atoms which act as bridges between two metal atoms. As expected from simple electrostatics and Eq. (9), bridging chlorines have a greater core binding energy than terminal chlorines. In Fig. 1a is shown the chlorine $2p_{1/2,\ 3/2}$ spectrum of $Re_3Cl_9(pyz)_3$ (pyz = pyrazine), in which there are twice as many terminal chlorines as bridging chlorines. Deconvolution shows that the two sets of binding energies differ by *ca.* 1.5 eV and that the doublet at lower binding energy is twice as intense as that at higher binding energy. Replacement of half of the terminal chlorines by other groups, as in the polymeric pyridine complex $[Re_3Cl_6(py)_3]_n$, yields a spectrum (Fig. 1b) in which the peaks due to terminal and bridging chlorines have a 1 : 1 intensity ratio. The diethyldithiocarbamate complex $Re_3Cl_3(SCN)_3(S_2CNEt_2)_3$, in which all the terminal chlorines have been replaced, gives a spectrum showing only the doublet due to the bridging chlorines.

Sometimes a simple qualitative interpretation of core binding energy data is sufficient to give a definitive answer to a problem involving atomic charge. Such was the case in the study of the ClF molecule by Carroll and Thomas[43]. These investigators found that the fluorine core electrons are less bound in ClF than in F_2 and that the chlorine 2s electrons are more bound in ClF than in Cl_2. Now, in the case of a diatomic molecule it is certain that, in Eq. (9), the absolute magnitude of the kQ_A term will always be greater than that of the potential term (which in this case is simply $-Q_A/R$). Hence if we accept the validity of Eq. (9), we conclude that an increase in binding energy corresponds to an increase in positive charge and that a decrease in binding energy corresponds to an increase in negative charge. As we shall soon point out, this conclusion is valid only if the change in electronic relaxation energy associated with the core ionization is zero. In the case of the molecules under consideration this is probably a good approximation. Inasmuch as the atomic charges in F_2 and Cl_2 are zero, the binding energy data strongly indicate that, in ClF, the chlorine atom is positively charged and the fluorine atom is negatively charged. Thus the XPS data refute the work of Flygare *et al.*[44, 45] who on the basis of the molecular Zeeman effect in ClF, came to the remarkable conclusion that the dipole moment corresponds to a positive charge on the fluorine and a negative charge on the chlorine.

The oxidation state of an atom in a compound is a quantity which is calculated according to a simple arbitrary recipe. If this recipe is agreed upon and understood, there should never be any argument as to the oxidation state of a particular atom in a compound. On the other hand, the *charge* of an atom is a quantity which must be calculated theoretically or somehow derived from experimental data. It is quite possible for an atom of a given element in a series of compounds to have the same oxidation state, and yet to have a wide range of atomic charges. Clearly attempts to determine oxidation states directly from core binding energies are potentially very risky. As we have seen, the relationship between binding energies and atomic charges is fairly straightforward, but the latter are related to oxidation states in a way which can be quite complicated. For example, it has been shown[46] that the nickel $2p_{3/2}$ binding energies of the dicarbollide complexes of Ni (III) and Ni (IV) (in which the nickel atoms are coordinated to boron and carbon atoms of the polyhedral ligand $B_9C_2H_{11}{}^{2-}$) are lower than that of $[Ni(H_2O)_4Cl_2] \cdot 2\,H_2O$ [in which an Ni (II) atom is coordinated to relatively electronegative oxygen and chlorine atoms]. Nevertheless, various workers have achieved reliable correlations of binding energy with oxidation state by restricting themselves to compounds with similar structures. Thus Chatt *et al.*[47] showed that the rhenium $4f_{7/2}$ binding energies of $ReCl_2(PMe_2Ph)_4$, $ReCl_3(PMe_2Ph)_3$, and $ReCl_4(PMe_2Ph)_2$ are linearly related to the rhenium oxidation states. Schmidbaur *et al.*[48] showed that compounds of gold in the + 1, + 2, and + 3 oxidation states having similar structures show a similar correlation of the gold $4f_{7/2}$ binding energy with oxidation state. And Edwards[49] found that the copper $2p_{3/2}$ binding energies of copper (I) carboxylates fall in the range 931.4–932.4 eV, whereas the range for copper (II) carboxylates is 934.0–934.4 eV.

D. Quantum Mechanical Calculations

Schwartz[50, 51] pointed out that the binding energy of a core electron is essentially equal to the potential felt at the core due to the nuclear charge and all the other electrons in the system. Chemical shifts can be related to a "valence electron potential", Φ_{val}, defined as

$$\Phi_{val} = -\sum_{A} P_i \left\langle r^{-1} \right\rangle_i + \sum_{i \neq A} (Q_i/R_i)$$

where P_i is the population in the ith valence orbital of atom A, $\left\langle r^{-1} \right\rangle_i$ is the radial expectation value for the orbital, and the second summation has the same significance that it has in Eq. (9). The valence electron potentials can be readily calculated by a semiempirical MO method such as CNDO/2 or Extended Hückel. Binding energies are then correlated by the relation

$$E_B = c\Phi_{val} + l' \tag{10}$$

where c is an empirically evaluated parameter which helps to compensate for approximations of the semiempirical MO method used to evaluate Φ_{val}, and l' represents the binding energy of the atom stripped of all valence electrons.

Ab initio calculations, that is, nonempirical LCAO SCF MO calculations, have been used in several different ways to calculate core binding energies and their shifts. We have already discussed the use of *ab initio* methods to calculate chemical shifts using the equivalent cores approximation. Such calculations have the advantage that they avoid hole-state calculations. However many direct hole-state calculations have been carried out. Often these calculations involve the assumption Koopmans' theorem[52]; that is, ionization potentials are assumed to be the same as the one-electron orbital energies of the parent neutral molecules. In this approximation, electronic relaxation and differences in correlation and relativitistic energies of molecules and ions are ignored. Differences in correlation and relativistic energies are expected to be less than about 1 eV for light atoms. However, electronic relaxation energies are considerably greater and pose a problem if absolute binding energies are calculated. A procedure which is theoretically more rigorous than the assumption of Koopmans' theorem is that in which *ab initio* calculations are carried out on both the molecule and its core-ionized state. An absolute binding energy, including relaxation energy, is then obtained by taking the difference between the calculated energies for these species. This procedure is often called the ΔSCF method. The difference between an absolute binding energy calculated in this way and the value calculated assuming Koopmans' theorem is essentially the electronic relaxation energy, E_R:

$$E_B \approx E_B(\Delta SCF) = E_B(KT) - E_R$$

In Table 5 we give some data of Hillier *et al.*[53] to illustrate the magnitude of calculated relaxation energies and the general quality of results which are obtained from *ab initio* calculations. The data in Table 5 were obtained using

Table 5. Experimental and calculated binding energies[53]

	Experimental		Koopmans' Theorem		ΔSCF		Relaxn.
	E_B	ΔE_B	E_B	ΔE_B	E_B	ΔE_B	Energy
N 1s							
NH_3	405.6	0.0	422.2	0.0	405.3	0.0	16.9
N_2	409.9	4.3	428.3	6.1	411.6	6.3	16.7
$N\overset{*}{N}O$	412.5	6.9	432.2	10.0	413.0	7.7	19.2
$\overset{*}{N}NO$	408.5	2.9	428.5	6.3	410.0	4.7	18.5
CH_3CN	405.9	0.3	424.5	2.3	406.4	1.1	18.1
O 1s							
H_2O	539.4	0.0	558.6	0.0	539.3	0.0	19.3
O_2 $^4\Sigma$	543.1	3.7	563.8	5.2	543.5	4.2	20.3
O_2 $^2\Sigma$	544.2	4.8	565.2	6.6	544.2	4.9	21.0
CO	542.3	2.9	563.5	4.9	543.4	4.1	20.1
N_2O	541.2	1.8	562.2	3.6	541.2	1.9	21.0
CO_2	540.8	1.4	562.8	4.2	542.4	3.1	20.3
C_3O_2	539.7	0.3	562.9	4.3	540.6	1.3	22.3
H_2CO	538.5	−0.9	560.1	1.5	538.8	−0.5	21.3

contracted Gaussian type functions, the expansion parameters being determined by least-squares fit to double-zeta Slater-type orbitals. Several conclusions can be drawn. First, the Koopmans' theorem binding energies are considerably greater (by about 4%) than the experimental and ΔSCF values. Second, the ΔSCF binding energies are in good agreement with the experimental values, the average deviation being ± 0.7 eV. Third, the chemical shifts calculated by either the Koopmans' theorem or the ΔSCF method are in fair accord with experiment, the ΔSCF values being much better in this respect than the Koopmans' theorem values. And fourth, relaxation energies for a core ionization can vary from one compound to another by as much as 2 or 3 eV.

E. Electronic Relaxation Energy

The principal stumbling block in the use of chemical shifts to study the nature of chemical bonding is the fact that the electronic relaxation energy is a variable. In other words, XPS does not directly give information about the ground state of a molecule. Thus Koopmans' theorem treatments and the unmodified point charge potential model [Eq. (9)], which is based on the atomic charges of ground-state molecules, are in principle inadequate. Fortunately there are various ways of estimating relaxation energies or of avoiding their complications.

The electronic relaxation energy associated with the core ionization of a molecule can be divided into two parts[54]:

$$E_R = E_R^{contr} + E_R^{flow} \tag{11}$$

The term E_R^{contr} is the relaxation energy associated with the contraction of the local charge distribution on the atom which is undergoing core ionization. The term E_R^{flow} is the relaxation energy associated with the rest of the molecule, that is with the flow of electron density toward the core hole. It has been pointed out[54–57] that E_R^{contr} is approximately a linear function of atomic charge (the more negative the charge, the greater the relaxation energy). Hence Eq. (11) can be written as follows,

$$E_R = k'Q + E_R^{flow} + l_R \tag{12}$$

where k' is a negative constant and l_R is the relaxation energy due to orbital contraction around a neutral atom in the molecule. The simple point charge potential equation [Eq. (9)] ignores relaxation energy. However, if we account for relaxation energy by substracting the expression for E_R [as given by Eq. (12)] from the right-hand side of Eq. (9), we see that the terms $-k'Q$ and $-l_R$ can be combined with the corresponding terms kQ_A and l of Eq. (9), resulting in the expression

$$E_B(A) = (k-k')Q_A + \sum_{i \neq A} (Q_i/R_i) + (l - l_R) - E_R^{flow} \tag{13}$$

Obviously whenever binding energy data are fit to Eq. (9), the empirically determined values of k and l automatically take into account the "atomic" relaxation energy, E_R^{contr}. Equation 9 gives very good correlations for molecules which have similar structures, presumably because of the automatic accounting for E_R^{contr} and the fact that similar molecules have similar E_R^{flow} values and that $-E_R^{flow}$ is therefore absorbed into the constant l.

When the simple point charge potential equation [Eq. (9)] is used to correlate binding energies for a wide variety of molecules having different types of structures, large discrepancies are often found between the calculated and experimental binding energies. Usually these discrepancies can be rationalized in terms of unusually large or small values of E_R^{flow}. For example, the carbon 1s binding energy of CO is much higher than one would predict from carbon compound correlations that do not take relaxation energy into account. This discrepancy is believed to be due to an unusually low relaxation energy for CO. In CO, the carbon atom is attached to only one other atom, whereas in almost all its other compounds carbon is attached to at least two other atoms. (One would expect the electron flow during relaxation to increase with an increase in the number of bonds, or avenues for electron flow[58, 59]). As another example of a discrepancy due to neglect of relaxation energy, we can cite the oxygen 1s binding energies of ketones and similar carbonyl compounds, which are usually much lower than predicted[60]. These core ionizations are believed to be accompanied by an extraordinarily large amount of relaxation, due to a shift of π electron density from the groups attached to the carbonyl group to the oxygen atom:

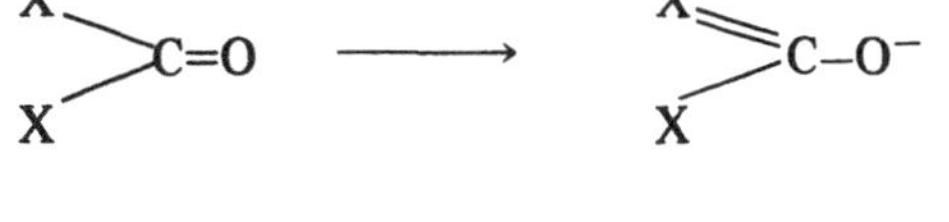

Such "π-donor relaxation" is of little importance in saturated oxygen compounds such as alcohols and ethers.

The point charge potential model [Eq. (9)] and the valence electron potential model [Eq. (10)] do not take account of electronic relaxation. They are based on the hypothetical "sudden" ejection of an electron, with the assumption that the energy of the process is dependent only on the initial ground-state charge distribution. However, data such as those in Table 5 and the success of the "equivalent cores" procedure (in which ground-state thermochemical data are used for excited states) indicate that the valence electron relaxation process is essentially complete in the time of the photoelectric process. Therefore if we wish to calculate accurately the binding energy by a hypothetical "sudden" process in which the valence electrons are assumed to remain fixed, we should not use the valence electron distribution of either the initial molecule or of the final core-hole ion. Use of the electron distribution of the initial molecule corresponds to the assumption of no relaxation energy (the absolute binding energy would be too high by an amount equal to the relaxation energy). Use of the electron distribution of the final ion corresponds to the inclusion of twice the appropriate relaxation energy (the absolute binding energy would be too low by an amount equal to the relaxation energy). However, a valence electron distribution between these two extremes, close to the average distribution, is appropriate[61)]. This procedure is equivalent to that achieved by quantum mechanical considerations by Liberman[62)] and Hedin and Johansson[63)] and is analogous to Slater's[64)] method for calculating excitation energies, in which one assumes occupation numbers half-way between those of the initial and final states. A logical name for the procedure of correlating binding energies with valence electron distributions half-way between those of the initial molecules and the core-ionized molecules is the "transition state" method. The equivalent cores approximation is used to calculate the electron distribution of the core-ionized molecule by simply replacing the core-ionized atom with the atom of the next element in the periodic table, plus a +1 charge. Alternatively, one can carry out the calculations directly for the transition state by using interpolated (pseudo-atom) calculational parameters (*e. g.*, CNDO) for the atom which undergoes core ionization. These methods, or very similar methods, have been applied very successfully to a wide variety of compounds using the valence electron potential model combined with CNDO calculations[58, 65)] and the point charge potential model combined with atomic charges estimated by the CHELEQ electronegativity equalization procedure[59, 66)] or the CNDO/2 method[67, 68)].

III. Shake-up

The X-ray photoelectron spectrum of the core ionization of an atom in a molecule consists of peaks and bands corresponding to transitions to various excited states. None of these transitions corresponds to the formation of the Koopmans' theorem frozen-orbital ionic state, which is a completely hypothetical state. However, the center of gravity of the various peaks and bands lies at the energy corresponding to the hypothetical Koopmans' theorem transition. This relation may be expressed by the sum rule of Manne and Aberg[69],

$$E_{KT} = (\sum_j E_j I_j)/(\sum_j I_j),$$

where E_{KT} is the Koopmans' theorem energy, and E_j and I_j are the energies and intensities of the various transitions. The primary peak, which is usually the most intense peak in the spectrum, occurs at lower binding energy than the Koopman's theorem energy; the energy difference corresponds to the electronic relaxation energy. In general, the spectrum also contains several satellite peaks with binding energies higher than that of the primary peak which correspond to electronically excited core-hole ion-molecules. These "shake-up" transitions may be looked upon as combinations of core ionizations and molecular electronic excitations. A broad band in the very high binding energy region of the spectrum corresponds to "shake-off", *i. e.*, to the loss of a valence electron as well as the core electron. For general discussions of the phenomena of shake-up and shake-off, the reader is referred to books by Siegbahn *et al.*[2] and Carlson[3] and a review article by Brisk and Baker[70].

If the primary peak is the only peak with a binding energy less than the Koopmans' theorem energy (this is usually the case), then an increase in the primary relaxation energy must be accompanied by an increase in the quantity $\sum_j (E_j - E_{KT}) I_j$ for the shake-up and shake-off bands, where the intensities I_j are understood to be measured relative to the primary peak. This function would increase, for example, if the relative intensity of any shake-up peak increased. As a matter of fact, it may be stated as a rough rule of thumb that core electron spectra for which the electronic relaxation energy is high are usually accompanied by intense shake-up peaks. This generalization implies that the shake-off bands are similar in both position and intensity and that the energies of the shake-up peaks are of similar magnitude. As an illustration of this rule of thumb we may compare the C 1s and O 1s spectra of CO with those of C_3O_2 and transition metal carbonyls. When we go from CO to the more complicated molecules, the number of atoms which can act as sources of valence electron density during the electronic relaxation processes increases, and therefore it is reasonable to suppose that the relaxation associated with both the C 1s and O 1s ionizations increases. The shake-up peaks in the CO spectra[71] are fairly weak, whereas those in the spectra of C_3O_2[72] and the metal carbonyls[73] are very intense[74, 75].

The selection rules appropriate for a shake-up transition are of the monopole type[2, 76]. The intensity of a shake-up peak depends on the overlap integral between the lower state molecular orbital from which the electron is excited (in the neutral molecule) and the upper state molecular orbital to which the electron is excited (in the core-ionized molecule). Consequently one expects transitions of the type $\sigma_u \rightarrow \sigma_u$, $\sigma_g \rightarrow \sigma_g$, $\pi_u \rightarrow \pi_u$, and $\pi_g \rightarrow \pi_g$ with $g \rightarrow u$ and $u \rightarrow g$ transitions forbidden.

As a first, trivial, example of the application of the overlap criterion, let us consider the possibility of a shake-up peak associated with the C 1s ionization of the terminal carbon atom in nitroethane and the $\pi \rightarrow \pi^*$ transition of the nitro group in that molecule. In this case the core ionization occurs in a region of the

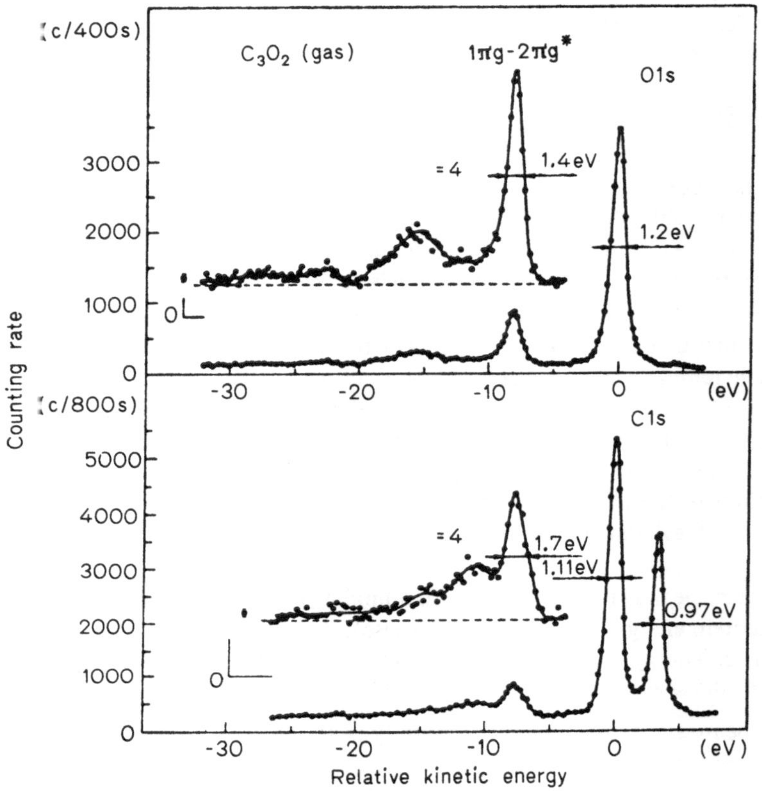

Fig. 2. The O 1s and C 1s regions of the X-ray photoelectron spectrum of C_3O_2, showing the shake-up satellites. Reproduced with permission from Ref.[77])

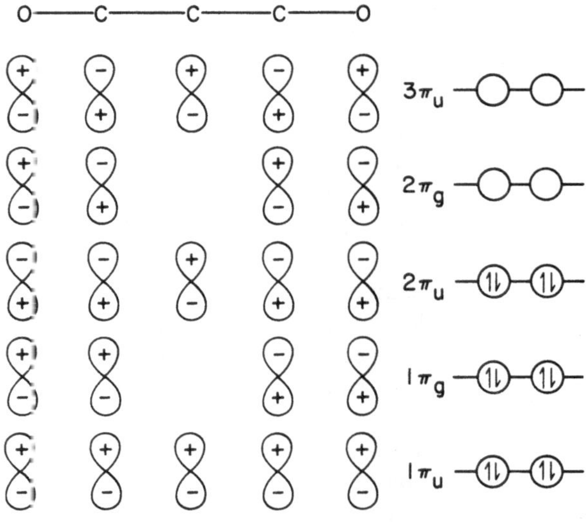

Fig. 3. The π molecular orbitals of C_3O_2. The atomic orbital contributions are shown only for one plane

molecule remote from the molecular orbitals involved in the electronic transition. Hence the upper-state molecular orbital will be essentially unperturbed by the core ionization and it will be, as in the neutral molecule, orthogonal with the lower-state molecular orbital. Consequently the critical overlap integral is almost zero and no shake-up is ovserved for that transition.

As a second example, let us consider the O 1s and C 1s spectra of carbon suboxide, C_3O_2, shown in Fig. 2. Strong satellites are seen 8.2 eV from the O 1s primary peak and about 7.9 eV from the more intense C 1s primary peak (the peak due the non-central carbon atoms). Gelius *et al.*[72, 77] believe that these satellites are due to the $1\pi_g \rightarrow 2\pi_g$ transition of C_3O_2, corresponding to the molecular orbitals schematically illustrated in Fig. 3. In the C 1s spectrum, no satellite is seen in the region about 8 eV from the less intense primary peak, *i. e.* the peak due to the central carbon atom. The lack of a satellite in the C 1s spectrum of the central carbon atom would follow from the fact that neither the $1\pi_g$ nor the $2\pi_g$ molecular orbital has any magnitude at the central carbon atom. If the interpretation is correct, this is a case in which the overlap integral between the lower-state and upper-state molecular orbitals is exactly zero. On the other hand, Aarons *et al.*[74] believe that the 8.2 and 7.9 eV satellites of the O 1s and non-central C 1s lines are due to the $2\pi_u \rightarrow 2\pi_g$ transition of C_3O_2. When one of the oxygen atoms or non-central carbon atoms in the $2\pi_g$ state of C_3O_2 is core-ionized, the state loses its *g* character and is no longer orthogonal with *u* states. Hence the latter assignment does not violate the $u \nleftrightarrow g$ selection rule. However, when the central carbon atom in the $2\pi_g$ state is core-ionized, the state retains its *g* character and the shake-up transition is forbidden.

Pignataro[78, 79] has observed bands centered about 5.5 eV from the main peaks in the C 1s and O 1s spectra of $Cr(CO)_6$.He has suggested that these are shake-up satellites corresponding to the ${}^1A_{1g} \rightarrow d^1\,T_{1u}$ metal→ligand charge transfer transition which appears at 5.5 eV in the ultraviolet absorption spectrum[80] of $Cr(CO)_6$ vapor. If this assignment is correct, it is another example in which the symmetry of a molecule is changed by a core-ionization, thus allowing an otherwise symmetry-forbidden transition to take place. A very weak satellite was observed about 5.8 eV from the primary peak in the Cr 3p peak of $Cr(CO)_6$. This peak is probably not a ${}^1A_{1g} \rightarrow d^1\,T_{1u}$ shake-up satellite because, in principle, such a satellite is strictly forbidden.

In general, the energy separation between a shake-up peak and the primary peak is not exactly equal to the corresponding electronic transition energy for the neutral molecule. And, in cases where a given electronic transition is represented by shake-up peaks in several core ionization spectra, the energy separations are usually slightly different. These differences can give information regarding the atomic orbital contributions to the relevant molecular orbitals.

The core ionization of an atom stabilizes all the valence electrons in the atom. Depending on whether the electronic transition shifts electron density to or from an atom, the energy separation for a shake-up peak of that atom will be less than or greater than the energy of the neutral molecule ionization[81]. As an illustration of these effects, let us consider the shake-up spectra of formamide, H_2NCHO[82]. The principal transitions involved are the $\pi_1 \rightarrow \pi_3{}^*$ and $\pi_2 \rightarrow \pi_3{}^*$ transitions. The π_1

molecular orbital consists of roughly equal amounts of carbon, nitrogen, and oxygen character, the π_2 molecular orbital has major contributions only from the nitrogen and oxygen atoms, and the $\pi_3{}^*$ molecular orbital has about twice as much carbon character as nitrogen or oxygen character. From the experimental data, given in Table 6, it can be seen that the energy of the band assigned to the $\pi_1 \rightarrow \pi_3{}^*$ transition is much lower in the carbon spectrum than in the nitrogen and oxygen spectra. This result is expected because the concentration of $\pi_3{}^*$ on carbon implies that a 1s hole on carbon stabilizes this orbital more than a 1s hole on nitrogen or

Table 6. Shake-up in the 1s spectra of formamide[73)]

Transition	C 1s ΔE	Intensity	N 1s ΔE	Intensity	O 1s ΔE	Intensity
Primary Peak	0	100	0	100	0	100
$\pi_2 \rightarrow \pi_3{}^*$	5?	< 0.5	9.9	3.5	7.4	7.3
$\pi_1 \rightarrow \pi_3$	9.6	7.2	15.0	5.3	13.9	8.5

oxygen would. The same sort of trend is expected in the case of the $\pi_2 \rightarrow \pi_3{}^*$ peaks, but in this case the carbon peak is extremely weak (as expected, because of the low π_2 density on the carbon atom) and there is considerable uncertainty as to the energy.

Relatively strong satellites, 5–10 eV from the main peaks, have been observed in the metal 2p spectra of 3d transition metal oxides and halides. These satellites

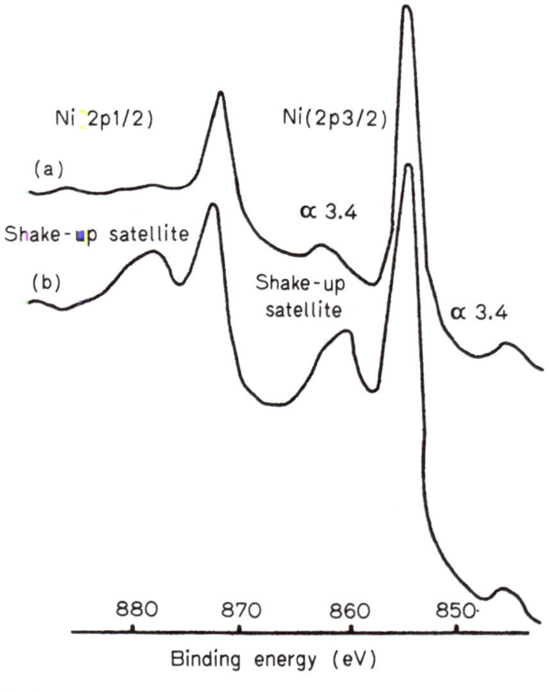

Fig. 4. The Ni 2p regions of the X-ray photoelectron spectra of the isomeric forms of the isomeric forms of bis(N-methylsalicylaldimine)nickel (II). A, square planar isomer; B, octahedral polymer. Reproduced with permission from Ref.[35)]

are believed to correspond to ligand → (metal 3d) charge-transfer transitions[76, 83–85].
Evidence for this assignment is found in the following facts.

1. Little satellite structure is found in the spectra of analogous cyanide complexes, except at excitation energies of 12–16 eV. The absence of high-intensity, low-energy shake-up peaks in the cyanide complexes is probably due to the fact that cyanide electrons are tightly bound and localized between the carbon and nitrogen atoms. The energy differences between metal 3 d levels and filled CN^- levels are large.
2. Satellite structure is generally most intense for paramagnetic compounds and for compounds of weak-field ligands. In such cases the energy separation between the ligand and metal orbitals is likely to be small. A nice illustration of this phenomenon is shown by the Ni 2p spectra of the two isomers of bis(N-methylsalicylaldimine)nickel (II), given in Fig. 4[35]. The diamagnetic square-planar isomer shows practically no evidence of shake-up, whereas the paramagnetic octahedral isomer shows strong shake-up satellits.
3. As one goes to metals of higher Z, the intensities of the shake-up peaks increase.
4. Although shake-up peaks are present in the case of metal ions with vacant or partially filled 3d shells, they are absent in the case of ions with completely filled 3d shells.

Finally, we point out that occasionally one observes spectra in which the intensities of the shake-up lines are comparable to, or even greater than, those of the primary lines. For example, such was the case in the Fe 2p spectrum of a sample of $KFeS_2$, obtained by Binder[86].

IV. Multiplet Splitting

Core electron ejection normally yields only one primary final state (aside from shake-up and shake-off states). However, if there are unpaired valence electrons, more than one final state can be formed because exchange interaction affects the spin-up and spin-down electrons differently. If a core s electron is ejected, two final states are formed. If a core electron of higher angular momentum, such as a 2p electron, is ejected, a large number of multiplet states can result. In this case it is difficult to resolve the separate states, and the usual effect of unpaired valence electrons is

1. to broaden $2p_{1/2}$ and $2p_{3/2}$ peaks, and
2. to increase the energy separation between these peaks[76].

The effect of spin on the width of the $2p_{3/2}$ line was studied by Main and Marshall[87]. In a survey of 26 different 3d transition metal compounds, they found that all 14 diamagnetic compounds (including ions of d^o, d^6 low spin, and d^{10} configurations) had $2d_{3/2}$ linewidths smaller than 2.5 eV and that 9 of the 12 paramagnetic compounds had linewidths greater than 2.5 eV. Several investigators have studied the effect of spin on the $2p_{1/2} - 2p_{3/2}$ separation[88–90]. It has been observed that the separation is 9.0–9.3 eV for most diamagnetic chromium compounds, 9.4–10.2 eV for most chromium(III) compounds[88], 15.1 ± 0.3 eV for diamagnetic cobalt (III) complexes and 16.0 ± 0.3 eV for cobalt (II) complexes[89, 90].

Small perturbations of the $2p_{1/2}-2p_{3/2}$ separations in the spectra of paramagnetic 3d transition metal compounds can be interpreted in terms of changes in the covalency of the metal ligand bonds. A decrease in the exchange interaction (leading to a decrease in the multiplet splitting or broadening and a decrease in the $2p_{1/2}-2p_{3/2}$ separation) is attributed to a delocalization of the unpaired electron density from the metal atom to the ligand atoms. For example, consider the Cr 2p data in Table 7[88]. In the complexes with "hard" ligands such as oxide and fluoride, the spin-orbit splitting is distinctly higher than it is in complexes with "soft" ligands such as urea and thiocyanate. The splitting approximately follows the nephelauxetic series of ligands[91], and it appears that in complexes such as $K_3[Cr(NCS)_6]$ the metal d orbitals are sufficiently expanded that they interact relatively weakly with the 2p core electrons. It has been reported that the multiplet splitting of the Cr 3s line in $K_3[Cr(CNS)_6]$ is intermediate between the values for $CrCl_3$ and $CrBr_3$[92].

Table 7. Spin-orbit separation for some chromium (III) compounds[88]

Compound	$2p_{1/2}-2p_{3/2}$ sepn., eV
K_3CrF_6	10.2
$NaCrO_2$	9.9
Cr_2O_3	9.7
$Cr(NH_3)_6Cl_3$	9.7
$CrCl_3$	9.6
$Cr(urea)_6Cl_3$	9.4
$K_3[Cr(NCS)_6]$	9.2

Because of the $1/r$ dependence, the exchange interaction rapidly decreases with an increase in the difference between the principal quantum numbers of the impaired electrons and the core vacancy. Thus multiplet splitting is negligible in the 1s spectra of paramagnetic transition metal compounds. On the other hand, shake-up intensity is essentially independent of the core level involved. Hence one can distinguish between shake-up and multiplet satellites by changing the principal quantum number of the core electron ionized[76].

V. Some Studies of Metal-Ligand Interactions

A. Nitrosyl Complexes

Nitrosyl complexes are of two types, depending on the orientation of the NO group with respect to the metal atom. In a "linear" complx, the M–N–O bond angle is near 180°, and the ligand is often considered to be the NO^+ ion; in a "bent" complex, the M–N–O bond angle is around 125°, and the ligand is often looked upon as the NO^- ion. In nitrosyl complexes it is obviously reasonable to look for a correlation between the N 1s binding energy and the type of coordination. Unfortunately there is only a very weak correlation, if any, of this type, although there is a strong correlation between the electron density on the metal atom and the N 1s binding energy[93, 94]. The N 1s binding energy is relatively low (in the range 399–401 eV) when the metal is in an extraordinarily low oxidation state, and relatively high (in the range 401–403 eV) when the metal is in a normal or high oxidation state. For example, the N 1s binding energies of $Ru(PPh_3)_3H(NO)$ and $Ru(PPh_3)_2Cl_3(NO)$ are 398.9 and 401.4 eV, respectively, although the NO groups are linear in both complexes[94]. Su and Faller found that the *difference* between the O 1s and N 1s binding energies shows a remarkable correlation with the orientation of the NO group[94]. In the case of linear nitrosyls, the O 1s – N 1s difference is 132 ± 2 eV; in the case of bent nitrosyls the difference is 128 ± 2 eV. The direction of this effect (*i. e.*, the fact that the difference for the linear nitrosyls is greater than that for the bent nitrosyls) does not seem to have a simple explanation. Perhaps the effect is due to a much greater relaxation energy associated with O 1 s ionization in the bent nitrosyls than in the linear nitrosyls.

B. Carbon Monoxide as a Ligand

1. Metal Carbonyls

The transition metal carbonyls represent a large and important class of compound. In a series of these compounds, the interaction of the carbon monoxide ligand with the metal atom can vary significantly. One cause for this variation is change in the degree of back-bonding, *i. e.*, in the extend to which $d\pi$ electron density is transferred from the metal atom to the oxygen atom of the CO ligand:

$$^{-}M{-}C{\equiv}O^{+} \leftarrow\rightarrow M{=}C{=}O$$

Because of the changes in formal charge which accompany back-bonding, one would expect the O 1s binding energy to be correlated with back-bonding. With this in mind, we have listed, in Table 8, the O 1s binding energies of a variety of metal carbonyls, all determined in the gas phase[95]. As an aid in looking for a correlation, we have also listed two other quantities which are expected to be functions of the degree of back-bonding in the CO ligand. One is the multiplicity-weighted average of the C–O stretching frequencies, $\langle \nu_{CO} \rangle$, listed for each compound for which all the required frequencies are known[96–108]. The quantity $\langle \nu_{CO} \rangle$ is expected to decrease with increasing back-bonding. The other quantity is the sum of n_V (the number of valence electrons in the free metal atom) and n_{CO} (the number of

Table 8. Correlation of oxygen 1 s binding energies of carbonyl complexes with back-bonding

Carbonyl Cpd.	E_B(O 1s), eV	ν_{CO}, cm^{-1}	Ref.	$(n_V + n_{CO})$
$C_6H_6Cr(CO)_3$	538.23	1938	96)	10
$C_7H_7V(CO)_3$	538.6	1935	97)	10
$C_5H_5V(CO)_4$	538.7	1963	98)	10
$C_5H_5(CH_3)Mo(CO)_3$	538.8	1971	99)	10
$C_5H_5Mn(CO)_3$	538.85	1974	100)	11
$Mn_2(CO)_{10}$	538.89	2017	100)	12
$C_7H_8Fe(CO)_3$	539.09	2004	101)	12
$CH_3(OCH_3)CCr(CO)_5$	539.1	1984	102)	12
$C_4H_6Fe(CO)_3$	539.29	2005	103)	12
$W(CO)_6$	539.52	2017	100)	12
$Mo(CO)_6$	539.61	2021	100)	12
$CH_3Re(CO)_5$	539.64	2033	104)	12
$Cr(CO)_6$	539.66	2018	100)	12
$HMn(CO)_5$	539.84	2039	105)	12
$CH_3Mn(CO)_5$	539.87	2032	106)	12
$V(CO)_6$	539.9			11
$CH_3COMn(CO)_5$	539.92	2038	107)	12.5
$Fe(CO)_5$	540.01	2035	100)	13
$Ni(CO)_4$	540.11	2066	100)	14
$Cl_3SiMn(CO)_5$	540.31	2058	108)	13

CO ligands and their equivalent[a] per metal atom in the compound). The quantity n_V changes in much the same way as the effective nuclear charge on the metal atom and, other things being equal, would be expected to be inversely related to the dπ donor ability of the atom. The quantity n_{CO} is a measure of the competition among ligands for dπ electron density and also would be expected to be inversely related to the dπ donor ability of the metal atom in the compound. Hence the sum $(n_V + n_{CO})$ should decrease with increasing back-bonding.

From Table 8 it can be seen that the three quantities E_B(O 1s), $\langle \nu_{CO} \rangle$, and $(n_V + n_{CO})$ are closely correlated. We conclude that the O 1s binding energy of a carbonyl can be used to estimate the degree of d$\pi \rightarrow$ pπ back-bonding: the lower the binding energy, the lower the degree of back-bonding. The one compound which appears to be out of line with respect to the correlation of E_B(O 1s) with $(n_V + n_{CO})$ is $V(CO)_6$. Perhaps the high binding energy in this case is due to the fact that the vanadium atom is one electron short of the krypton effective atomic number. This electron deficiency would be expected to make the vanadium atom a relatively poor dπ donor.

[a] In the case of carbonyls containing ligands other than CO, it was assumed that the following ligands are equivalent, as π acceptors, to the indicated number of CO groups: C_7H_7, 2; C_5H_5, C_6H_6, C_4H_6, C_7H_8 (tetrahapto), $SiCl_3$, $CH_3(CH_3O)C$, CH_3CO, 0.5. Although these assumptions are somewhat *ad hoc*, it can be seen that they are chemically reasonable.

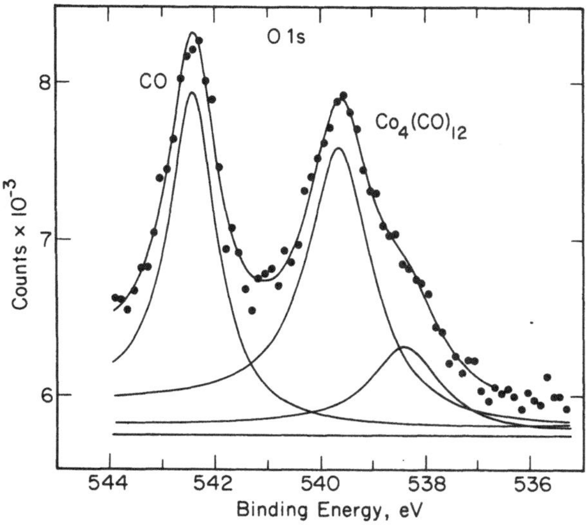

Fig. 5. The O 1s spectrum of gaseous $Co_4(CO)_{12}$. The left-hand peak is due to CO formed in the decomposition of the $Co_4(CO)_{12}$. The $Co_4(CO)_{12}$ band has been deconvoluted into two peaks with relative areas of 3:1. Reproduced with permission from Ref.[109]

There are two principal modes of coordination of the carbon monoxide ligand in transition metal carbonyls: terminal coordination, to a single metal atom, and bridging coordination, to two or more metal atoms. The O 1s spectrum of $Co_4(CO)_{12}$, shown in Fig. 5, can be readily deconvoluted into two peaks corresponding to these two types of carbonyl groups[109]. This spectrum is useful for determining the relative chemical shifts for the two types because

1. the groups are attached to identical cobalt atoms, and thus shifts due to changes in the metal and other ligands are avoided, and
2. the $Co_4(CO)_{12}$ molecule contains terminal and bridging carbonyl groups in a 3:1 abundance ratio, and thus the deconvoluted peaks can be readily assigned on the basis of their relative intensities.

From the spectrum one concludes that the O 1s binding energy of the terminal carbonyl group is 1.25 eV higher than that of the bridging carbonyl group. This result is qualitatively in accord with what one might expect, assuming greater back-bonding in the bridging carbonyl groups. It is also the result expected on the basis of simple valence bond formulas in which the terminal carbonyl group is represented with no back-bonding and the bridging carbonyl group is represented by a simple keto structure.

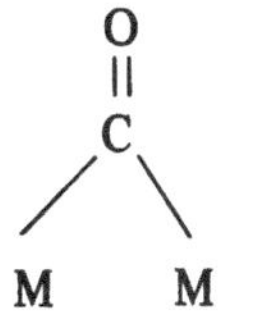

2. Adsorption of CO on Metal Surfaces

The chemisorption of carbon monoxide on a metal surface involves metal-ligand interactions which are similar to those in metal carbonyls as well as interactions which are peculiar to the surface. For example, Yates, Madey, and Erickson[110, 111] have interpreted the O 1s spectrum of CO adsorbed on tungsten in terms of six different types of adsorbed carbon monxide: "virgin," α_1, α_2, β_1, β_2 and β_3. The virgin CO, for which E_B(O 1s) = 536.0 eV, is identified with carbon monoxide coordinated to two or more metal atoms and is structurally analogous to bridging CO groups in molecular carbonyls. This species is formed during adsorption at low temperatures and low surface coverage. The α_1 and α_2 forms [E_B(O 1s) = 538.7 and 537.3 eV, resp.] are identified with weakly-bound carbon monoxide, coordinated through the carbon atom to only one metal atom. The α_2 form is supposed to have more back-bonding from the tungsten than the α_1 form. Presumably this difference in the nature of the metal-ligand bonding arises from adsorption on different surface sites, such as terrace sites (where the metal d electrons are delocalized and relatively unavailable for back-bonding) and step sites (where the metal d electrons are localized and relatively available for back-bonding). It is significant that the O 1s spectrum of CO adsorbed on a tungsten (100) single crystal surface could not be readily resolved into more than one peak in the α-CO region, whereas the spectrum of CO adsorbed on polycrystalline tungsten was deconvoluted to give α_1 and α_2 peaks. The binding energies of the α forms and virgin forms are qualitatively in agreement with the values found for the analogous terminal and bridging carbonyl groups of molecular carbonyls. The β_1, β_2, and β_3 forms [E_B(O 1s) in the range 534.5–537.1 eV] are identified with carbon monoxide in which both the carbon and oxygen atoms are directly bonded to the tungsten metal. The E_B(O 1s) of the β_2 form (~ 534.5 eV) is close to that observed for species which form in the adsorption of O_2 and CO_2 on tungsten. It has been suggested that these coincidences in O 1s binding energy reflect a limiting condition for the oxygen of CO, O_2, and CO_2. Perhaps in each case complete molecular fragmentation (dissociative sorption) occurs to form chemisorbed oxygen atoms, with E_B(O 1s) $\approx$ 534.5 eV.

Some interesting results have been obtained in XPS studies of the adsorption of CO on a single crystal surface on which 5/6 of the surface platinum atoms were on terraces (the 111 plane) and 1/6 of the surface atoms were at step and kink sites, as shown in Fig. 6[112]. The initial adsorption of carbon monoxide was dissociative, as evidenced by the appearance in the C 1s spectrum of a "carbide" or "carbon atom" peak with a binding energy of 283.8 eV. The intensity of the C 1s peak corresponded to that expected for quantitative adsorption at the step and kink sites. The ratio of the C 1s to O 1s peak intensities was slightly greater than expected for a 1:1 ratio of C and O atoms, suggesting that some of the oxygen atoms reacted with CO to form gaseous CO_2. Increasing exposure of the surface to CO caused the growth of a second C 1s peak with a binding energy of 286.7 eV. This binding energy is equal to that found for CO adsorbed on an unstepped Pt (111) surface and corresponds to associatively adsorbed CO groups on the terraces. The fact that associative adsorption occurs on platinum terraces and that dissociative adsorption occurs at platinum steps and kinks may be related to the relative electron densities at these sites. At steps

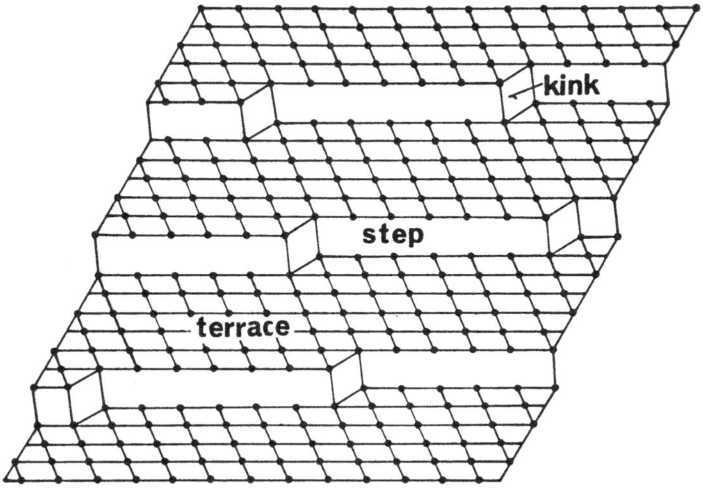

Fig. 6. A model of a platinum surface on which 5/6 of the atoms are on terraces (the 111 plane) and 1/6 of the atoms are at step and kink sites. Reproduced with permission from Ref.[112])

and kinks, the electron density is presumably high enough to cause effective reduction of the atoms to "carbide" and "oxide" ions.

Exposure of the clean stepped platinum surface to oxygen caused saturation of the step and kink sites (no adsorption occurred on a 111 surface under identical conditions). The oxygen atom-saturated surface was then exposed to varying amounts of carbon monoxide. Both "carbide" carbon and CO carbon C 1s peaks formed, with a one-to-one correspondence between the growth of carbide and the decrease of surface oxygen atoms. These data are consistent with threee possible reaction schemes:

$$O(ad) + CO(ad) \longrightarrow CO_2(g) \qquad (i)$$

$$O_2(g) + CO(ad) \longrightarrow CO_2(g) \qquad (ii)$$

$$O(ad) + CO(g) \longrightarrow CO_2(g) \qquad (iii)$$

The first scheme was ruled out by showing that, at room temperature, a surface formed by very brief exposure of the oxygen-saturated surface to carbon monoxide is stable after removal of the carbon monoxide from the reaction chamber. In other words, no further surface carbide formed by lateral reactions of adsorbed carbon monoxide with surface oxygen atoms. The second scheme was ruled out by showing that exposure of the surface formed in the latter experiment to oxygen had no effect. Consequently the third scheme is believed to represent the mechanism of oxidation of carbon monoxide at the step and kink sites of platinum.

C. Olefins as Ligands

1. Olefin Complexes

The bonding of an olefin to a metal atom is usually described as a synergistic combination of σ donor bonding and π back-bonding. Studies of the metal core binding energies of olefin complexes are consistent with this bonding description; indeed, the results suggest that more charge is transferred in the back-bonding than in the donor bonding. Thus Cook *et al.*[113] have shown that the Pt $4f_{7/2}$ binding energy of $(C_2F_4)Pt(PPh_3)_2$ is 0.6 eV higher than that of $Pt(PPh_3)_4$. By making the reasonable assumption that the platinum atom in the tetrakis(phosphine) complex has a zero charge, they concluded that the platinum atom in the olefin complex is positively charged. Zakharova *et al.*[114] found that, on going from K_2MCl_4 to the 1,5-hexadiene complex $C_6H_{10}MCl_2$, the metal core binding energy increased by 0.4 eV for M = Pd and by 1.5 eV for M = Pt. The greater increase in the case of the platinum complexes was explained by the greater ability of the platinum atom to participate in back-bonding.

The Pt $4f_{7/2}$ binding energy changes by no more than 0.2 eV in the series of complexes $(C_2R_4)Pt(PPh_3)_2$, where R = H, Cl, F, Cn, and C_6H_5[113, 115]. This essential constancy of the platinum binding energy has been taken as evidence that the Lewis basicity of the bis(triphenylphosphine)platinum (O) moiety is so strong that the substituent groups on the olefin play a minor role in determining charge transfer from the metal to ligand antibonding orbitals. In a series of complexes (olefin)-$Ni[P(O\text{-}o\text{-}tolyl)_3]_2$ a slight trend in the Ni $2p_{3/2}$ binding energy (total change: 0.4 eV) was observed: C_2H_4 < acrylonitrile < maleic anhydride[116]. By contrast, the Ir $4f_{7/2}$ binding energy of olefin complexes of Vaska's complex, $Ir(CO)Cl(PPh_3)_2$, changed significantly on going from one olefin to another[115, 117]. For example, a 1.0 eV increase was observed on going from the C_2F_4 complex to the $C_2(CN)_4$ complex. It was inferred that Vaska's complex has a relatively low basicity and that metal to ligand charge transfer is significantly determined by the nature of the adduct.

2. Adsorption of Olefins on Metal Surfaces

Mason and co-workers have used XPS as an analytical tool for studying the reactions of haloalkenes on a platinum (100) surface[118–120]. In Fig. 7 the results of several sets of experiments are presented graphically[119]. The data show that the low-coverage chemisorptions of vinyl chloride and vinyl fluoride proceed without the appearance of any XPS peak for the halogen. The halogen peak appears only after about 0.4 monolayer coverage, and beyond that point the halogen and carbon peak intensities increase in proportion to the known relative cross-sections for these atoms. Similar results were found for 1,1-difluoroethylene. It is concluded that these molecules are dissociatively adsorbed at low coverages (with the loss of hydrogen halide to the gas phase), and associatively adsorbed at high coverages. The olefins 1,2-difluoroethylene and trifluorochloroethylene were found to adsorb associatively at all coverages. Evidence that the dehydrohalogenation involved in the dissociative sorption is

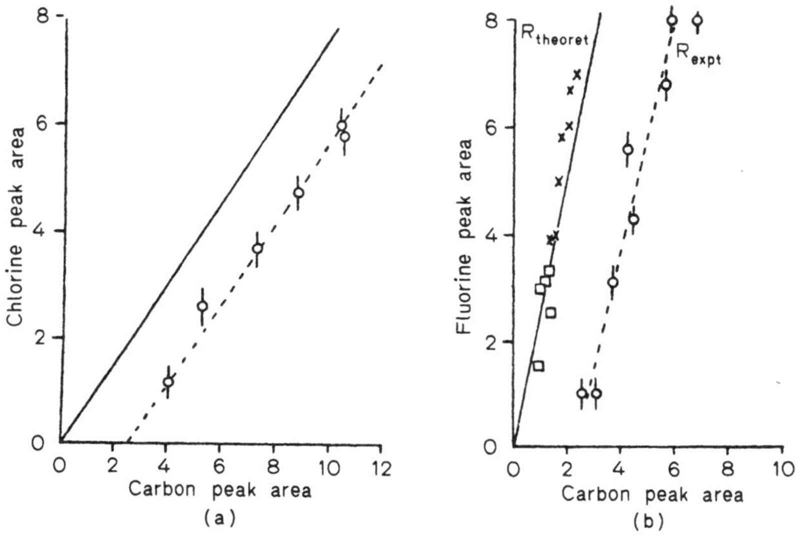

Fig. 7. Plots of halogen XPS peak area *vs.* carbon 1s peak area for vinyl halides adsorbed on a platinum 100 surface.
(a) Vinyl chloride. The solid line corresponds to the ratio (0.78) of the Cl 2p and C 1s emission intensities in the gas phase spectrum.
(b) Open circles correspond to vinyl fluoride data. The solid line corresponds to the ratio (2.27) of the F 1s and C 1s emission intensities in the gas phase spectrum. The crosses correspond to preadsorption of 0.3 monolayer of CO; the squares to preadsorption of 0.6 monolayer of CO. Reproduced with permission from Ref.[119]

intermolecular rather than intramolecular was found in a study of the sequential adsorption of two different olefins. When the initial associative adsorption of trifluorochloroethylene was followed by the adsorption of ethylene, dehydrohalogenation took place.

A mechanism which has been offered to explain these and other data is shown in the following scheme (s. p. 176). The initial step corresponds to the metallation of a metal-hydrogen bond. The rate of metallation of a haloalkene (which is essentially an electrophilic attack by the metal on the bond) depends on the electron density in the C–H bond. Obviously electron-withdrawing substituents such as halogen atoms would be expected to reduce the rate of metallation. We can rationalize the data for the reactions on platinum (100) by assuming that a halogen atom directly attached to the carbon of a C–H bond reduces the rate of metallation to a negligible value.

The change in the nature of the adsorption with increasing coverage (dissociative followed by associative) has been explained by a statistical consideration of the reaction mechanism shown above[120]. Associative adsorption is expected to occur at vacant sites for which all adjacent olefin binding sites are occupied by earlier dissociation products (or carbon monoxide, as shown by Fig. 6b), because dissociative adsorption (formation of vinyl and hydride species, followed by hydride migration to another alkene) requires two adjacent vacant sites.

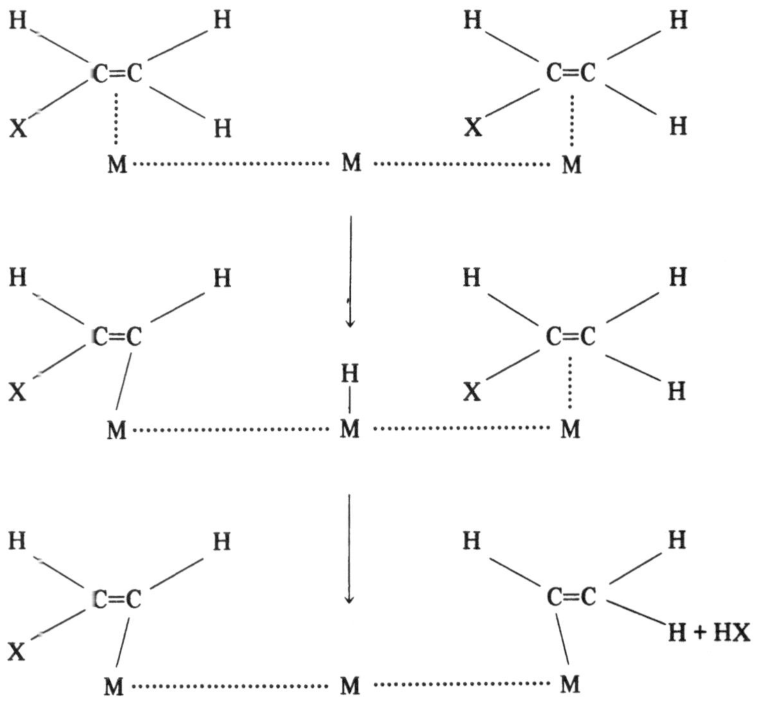

VI. Mixed Valence Compounds

For the purpose of this discussion, a mixed valence compound is defined as a compound containing two or more metal atoms with an average oxidation state which is either

1. non-integral or
2. unusual for the metal and intermediate between two well-known oxidation states of the metal.

Compounds of this type have received much study in recent years[121]. One of the important properties to be determined for a mixed valence compound is the extent of electronic coupling between the metal atoms. Do the metals atoms have the same, average, oxidation state (corresponding to a delocalization of the metal valence electrons), or do they have different oxidation states (corresponding to localized valence electrons)? Before answering this question, one must decide what lifetime for a particular valence electronic state is the upper limit for a delocalized system and is the lower limit for a localized system. Then, to answer the question, one must employ a physical measurement which has a time scale of the same magnitude as that of the chosen lifetime[122]. For example, if Mössbauer spectroscopy shows the presence of two different kinds of iron atoms in a mixed valence compound, one knows that the lifetime of the state must be longer than about

10^{-7} s. If infrared spectroscopy shows that the metal atoms of a mixed valence compound are structurally equivalent, one knows that the lifetime of any electronic state in which the atoms are nonequivalent must be less than about 10^{-13} s (the time required for a bond vibration). If the lifetime were longer, atoms would shift so as to trap the compound in the state with nonequivalent atoms.

X-Ray photoelectric ionization is believed to take place in a time interval of about 10^{-18} s. Therefore separate XPS peaks are possible for atoms if the lifetime of the asymmetric electronic state is greater than about 10^{-18} s, *whether or not the atoms are structurally equivalent.* We may represent the ground state of a localized mixed valence compound (involving two metal atoms differing in oxidation state by one unit) by the following formula, where the dot represents the extra valence electron: $\mathrm{M{-}\dot{M}}$. The two possible XPS transitions can then be represented as follows, where the asterisk indicates core ionization,

$$\mathrm{M{-}\dot{M}} \longrightarrow [\mathrm{M^{*}{-}\dot{M}}]^{+} + e^{-}$$
$$\mathrm{M{-}\dot{M}} \longrightarrow [\mathrm{M{-}\dot{M}^{*}}]^{+} + e^{-}$$

The first transition would be expected to be of higher energy than the second from simple atomic charge considerations. Because the two atoms are of equal abundance, the two peaks have essentially equal intensities. Unfortunately, the observation of two XPS peaks does not rule out the possibility of delocalized valence electrons in the ground state. Two transitions are expected even in that case because of polarization of the excited state by the core ionization[123)]. The ground state of a delocalized mixed valence compound can be crudely represented by the formula $\mathrm{M\dot{-}M}$, where the intermediate position of the dot indicates that the odd valence electron is equally shared by the two metal atoms. The two XPS transitions can then be represented as follows,

$$\mathrm{M\dot{-}M} \longrightarrow [\mathrm{M^{*}{-}\dot{M}}]^{+} + e^{-}$$
$$\mathrm{M\dot{-}M} \longrightarrow [\mathrm{M{-}\dot{M}^{*}}]^{+} + e^{-}$$

Notice that the excited states are similar to those formed from the localized ground states. Again the first transition is of higher energy than the second. Hush has made a theoretical study of these transitions[123)]. He concluded that, in the case of symmetrical delocalized mixed valence complexes, the two XPS peaks will occur at energies

$$1/2\,[E_{\mathrm{B}}(\mathrm{unocc}) + E_{\mathrm{B}}(\mathrm{occ})] + 1/2\,\Delta[\alpha \pm (1 + \alpha^2)^{1/2}]$$

where Δ is the difference between the orbital energies in the excited state and $\alpha = 2|J|/\Delta$, where J is the electronic interaction integral. He similarly concluded that the relative intensities of the two peaks will be

$$\frac{I(\text{low } E_B)}{I(\text{high } E_B)} = \left(\frac{1 + (1 + \alpha^2)^{1/2} + \alpha}{1 + (1 + \alpha^2)^{1/2} - \alpha}\right)^2$$

There may be some profit in looking upon the low- and high-energy transitions of a symmetrical delocalized complex as primary core ionization and shake-up transitions, respectively. According to this interpretation, the intensities of the two peaks will not, in general, be equal. The intensity of the "shake-up" peak would be expected to be proportional to the overlap between the wave functions of the metal valence electron in the delocalized ground state and in the excited, core-ionized state. The half-width of the "shake-up" peak would be expected to be greater than that of the "primary" peak because the excited core-ionized state would probably have a more strained geometry than the unexcited state.

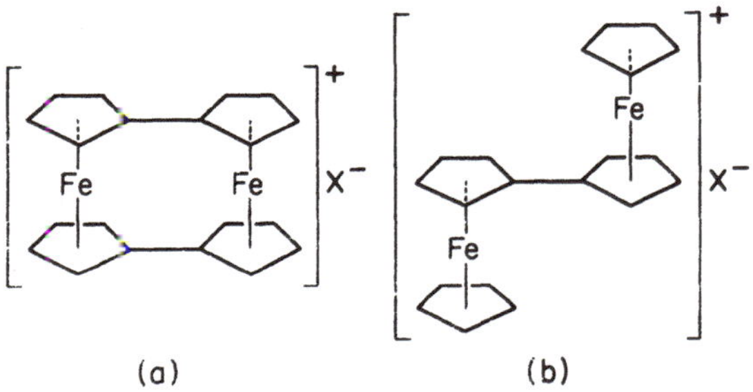

Fig. 8. Structures of the biferrocenylene (II, III) cation (a) and the biferrocene (II, III) cation (b). Reproduced with permission from Ref.[5)]

It must be admitted that, with our present understanding of these compounds and their core-ionization phenomena, the interpretation of two-peak XPS spectra for mixed valence compounds is necessarily somewhat ambiguous. Considerable further study will be required before we are able to extract the information inherent in such spectra. Nevertheless, let us consider several examples of 2-peak spectra which have been described in the literature. The Mössbauer spectrum of biferrocene (II, III) picrate (which has the structure given in Fig. 8) shows the existence of two distinctly different types of iron atoms, indicative of a localized ground state[124)]. The Fe $2p_{3/2}$ spectrum is consistent with this result; two peaks of approximately equal area, separated by 3.4 eV, are observed[124)]. The higher energy peak, corresponding to Fe (III), is much broader than the lower energy peak probably because of unresolved multiplet splitting. Crystal structure studies of Cs_2SbCl_6 have shown that the Sb (III) and Sb (V) atoms occupy sites which are slightly different. In accord with this result, the Sb $3d_{5/2}$ spectrum shows two peaks of equal intensity and half-width, separated by 1.8 eV[125)]. The physical properties of the ruthenium (II, III) complex ion $(NH_3)_5Ru(pyr)Ru(NH_3)_5{}^{5+}$ do not unequivocally indicate whether the ion is localized or delocalized[126, 127)] Citrin[128)] has observed two Ru $3d_{5/2}$ peaks, separated by 2.8 eV, for this complex. As we have pointed out, such a result does not prove that the extra Ru valence electron is localized. In fact, Hush[123, 127] has shown that Citrin's data, as well as the near-infrared electronic transition and the vibrational spectra, can be interpreted in terms of a delocalized complex.

The appearance of only one XPS peak for a mixed valence compound is consistent with a delocalized ground state (and excited state). Biferrocenylene (II, III) picrate, whose structure is shown in Fig. 8, probably fits in this category. The Mössbauer spectrum of the complex indicates only one kind of iron atom, and the Fe $2p_{3/2}$ spectrum consists of only one peak with a weak shoulder at higher binding energy[129]. It should be recognized, however, that even in the case of a localized system in which two XPS peaks are expected, if the chemical shift between the two peaks is less than the resolution of the spectrometer, only one peak will be observed.

Acknowledgements. This work was supported by the U.S. Energy Research and Development Administration and the National Science Foundation (Grant CHE 73-05133 AO2).

VII. References

1) Siegbahn, K., *et al.*: ESCA; atomic molecular and solid-state structure studied by means of electron spectroscopy. Uppsala: Almqvist and Wiksells AB 1967
2) Siegbahn, K., *et al.*: ESCA applied to free molecules. Amsterdam: North-Holland Publ. Co. 1969
3) Carlson, T. A.: Photoelectron and auger spectroscopy. New York: Plenum Press 1975
4) Hamnett, A., Orchard, A. F.: Electronic Structure and Magnetism of Inorganic Compounds *1*, 36 (1972); Evans, S., Orchard, A. F.: Electronic Structure and Magnetism of Inorganic Compounds *2*, 20 (1973); Hamnett, A., Orchard, A. F.: Electronic Structure and Magnetism of Inorganic Compounds *3*, 218 (1974)
5) Jolly, W. L.: Coord. Chem. Rev. *13*, 47 (1974)
6) Jolly, W. L. in Electron spectroscopy: Theory, techniques and applications. Brundle, C. R., Baker, A. D. (eds), Vol. I. London: Academic Press 1977, pp. 119–149
7) Hercules, D. M.: Anal. Chem. *44*, 106R (1972); Hercules, D. M., Carver, J. C.: Anal. Chem. *46*, 133R (1974); Hercules, D. M.: Anal. Chem. *48*, 294R (1976)
8) Corrected value from Perry, W. B., Jolly, W. L.: Inorg. Chem. *13*, 1211 (1974)
9) Holmes, S. A., Thomas, T. D., Unpublished work
10) Gelius, U., Héden, P. F., Hedman, J., Lindberg, B. J., Manne, R., Nordberg, R., Nordling, C., Siegbahn, K.: Physica Scripta *2*, 70 (1970)
11) Jolly, W. L.: J. Am. Chem. Soc. *92*, 3260 (1970)
12) Lindberg, B. J., Hedman, J.: Chemica Scripta *7*, 155 (1975)
13) Gray, R. C., Carver, J. C., Hercules, D. M.: J. Electron Spectr. Rel. Phen. *8*, 343 (1976)
14) Hedman, J., Klasson, M., Lindberg, B. J., Nordling, C., in: Electron spectroscopy (Shirley, D. A.)(ed.). Amsterdam: North-Holland Publ. Co. 1972, p. 813
15) Fluck, E., Weber, D.: Z. anorg. allgem. Chem. *412*, 47 (1975)
16) Fluck, E., Weber, D.: Pure Appl. Chem. *44*, 373 (1975)
17) Lindberg, B., Svensson, S., Malmqvist, P. A., Basilier, E., Gelius, U., Siegbahn, K.: Uppsala University Institute of Physics Report 910 (1975)
18) Jolly, W. L., Bakke, A. A.: J. Am. Chem. Soc. *98*, 6500 (1976)
19) Swain, C. G., Lupton, E. C.: J. Am. Chem. Soc. *90*, 4328 (1968)
20) Jolly, W. L., Hendrickson, D. N.: J. Am. Chem. Soc. *92*, 1863 (1970)
21) Hollander, J. M., Jolly, W. L.: Acc. Chem. Res. *3*, 193 (1970)
22) Finn, P., Pearson, R. K., Hollander J. M., Jolly, W. L.: Inorg. Chem. *10*, 378 (1971)
23) Finn, P., Jolly, W. L.: J. Am. Chem. Soc. *94*, 1540 (1972)
24) Jolly, W. L., in: Electron spectroscopy.(Shirley, D. A.) (ed.). Amsterdam: North-Holland Publ. Co. 1972, pp. 629–645
25) Thomas, T. D., J. Am. Chem. Soc. *92*, 4184 (1970)
26) Thomas, T. D., Shaw, Jr., R. W.: J. Electron Spectrosc. Relat. Phenom. *5*, 1081 (1974)
27) Davis, D. W., Hollander, J. M., Shirley, D. A., Thomas, T. D.: J. Chem. Phys. *52*, 3295 (1970)
28) Cavell, R. G.: J. Electron Spectrosc. Relat. Phenom. *6*, 281 (1975)
29) Clark, D. T., Adams, D. B.: Nature Phys. Sci. *234*, 95 (1971)
30) Frost, D. C., Herring, F. G., McDowell, C. A., Woolsey, I. S.: Chem. Phys. Letters *13*, 391 (1972)
31) Dewar, M. J. S., Lo, D. H.: Chem. Phys. Letters *33*, 298 (1975)
32) Adams, D. B., Clark, D. T.: Theoret. Chim. Acta *31*, 171 (1973)
33) Aarons, L. J., Hillier, I. H.: J. Chem. Soc. Faraday II *69*, 1510 (1973)
34) Pauling, L.: The nature of the chemical bond, 3rd edit. Ithaca, New York: Cornell Univ. Press 1960; also see 2nd edit. 1940, pp. 65–66
35) Matienzo, L. J., Yin, L. I., Grim, S. O., Swartz, W. E.: Inorg. Chem. *12*, 2762 (1973)
36) Grim, S. O., Matienzo, L. J.: Inorg. Chem. *14*, 1014 (1975)
37) Hughes, W. B., Baldwin, B. A.: Inorg. Chem. *13*, 1531 (1974)
38) Moddeman, W. E., Blackburn, J. R., Kumar, G., Morgan, K. A., Jones, M. M., Albridge, R. G., in: Electron spectroscopy. (Shirley, D. A.)(ed.). Amsterdam: North-Holland Publ. Co. 1972, pp. 725–32

39) Swartz, W. E., Ruff, J. K., Hercules, D. M.: J. Am. Chem. Soc. *94*, 5227 (1972)
40) Citrin, P. H., Shaw, R. W., Packer, A., Thomas, T. D., in: Electron spectroscopy. (Shirley, D. A.) (ed.). Amsterdam: North-Holland Publ. Co. 1972, pp. 691–706; also see J. Chem. Phys. *57*, 4446 (1972)
41) Parry, D. E.: J. Chem. Soc. Faraday II *71*, 337 (1975)
42) Hamer, A. D., Walton, R. A.: Inorg. Chem. *13*, 1446 (1974)
43) Carroll, T. X., Thomas, T. D.: J. Chem. Phys. *60*, 2186 (1974)
44) Ewing, J. J., Tigelaar, H. L., Flygare, W. H.: J. Chem. Phys. *56*, 1957 (1972)
45) McGurk, J., Norris, C. L., Tigelaar, H. L., Flygare, W. H.: J. Chem. Phys. *58*, 3118 (1973)
46) Pont, L. O., Siedle, A. R., Lazarus, M. S., Jolly, W. L.: Inorg. Chem. *13*, 483 (1974)
47) Chatt, J., Elson, C. M., Hooper, N. E., Leigh, G. J.: J. Chem. Soc. Dalton *1975*, 2392
48) Schmidbaur, H., Mandl, J. R., Wagner, F. E., Van de Vondel, D. F., Van der Kelen, G. P.: J. Chem. Soc. Chem. Commun. *1976*, 170
49) Edwards, D. A.: Inorganica Chimica Acta *18*, 65 (1976)
50) Schwartz, M. E.: Chem. Phys. Lett. *6*, 631 (1970)
51) Schwartz, M. E., in: Electron spectroscopy. (Shirley, D. A.) (ed.).Amsterdam: North-Holland Publ. Co. 1972, p. 605
52) Koopmans, T. A.: Physica *1*, 104 (1933)
53) Aarons, L. J., Guest, M. F., Hall, M. B., Hillier, I. H.: J. Chem. Soc. Faraday II *69*, 563 (1973)
54) Gelius, U., Siegbahn, K.: Disc. Faraday Soc. Chem. Soc. *54*, 257 (1972)
55) Snyder, L. C.: J. Chem. Phys. *55*, 95 (1971)
56) Guest, M. F., Hillier, I. H., Saunders, V. R., Wood, M. H.: Proc. Roy. Soc. London (A) *333*, 201 (1973)
57) Clark, D. T., Scanlan, I. W., Muller, J.: Theoret. Chim. Acta (Berl.) *35*, 341 (1974)
58) Davis, D. W., Shirley, D. A.: Chem. Phys. Lett. *15*, 185 (1972)
59) Jolly, W. L., Perry, W. B.: J. Am. Chem. Soc. *95*, 5442 (1973)
60) Jolly, W. L., Schaaf, T. F.: J. Am. Chem. Soc. *98*, 3178 (1976)
61) Jolly, W. L.: Discuss. Faraday Soc. *54*, 13 (1972)
62) Liberman, D.: Bull. Amer. Phys. Soc. *9*, 731 (1964)
63) Hedin, L., Johansson, A.: J. Phys. Soc. London (At. Mol. Phys.) *2*, 1336 (1969)
64) Slater, J. C.: Advan. Quantum Chem. *6*, 30 (1972)
65) Davis, D. W., Shirley, D. A.: J. Electron Spectr. Rel. Phen. *3*, 137 (1974)
66) Jolly, W. L., Perry, W. B.: Inorg. Chem. *13*, 2686 (1974)
67) Howat, G., Goscinski, O.: Chem. Phys. Lett. *30*, 87 (1975)
68) Siegbahn, H., Medeiros, R., Goscinski, O.: J. Electron Spectr. Rel. Phen. *8*, 149 (1976)
69) Manne, R., Aberg, T.: Chem. Phys. Letters *7*, 282 (1970)
70) Brisk, M. A., Baker, A. D.: J. Electron Spectrosc. Rel. Phen. *7*, 197 (1975)
71) Gelius, U.: J. Electron Spectrosc. Rel. Phen. *5*, 985 (1974)
72) Gelius, U., Allen, C. J., Allison, D. A., Siegbahn, H., Siegbahn, K.: Chem. Phys. Letters *11*, 224 (1971)
73) Barber, M., Connor, J. A., Hillier, I. H.: Chem. Phys. Letters *9*, 570 (1971)
74) Aarons, L. J., Guest, M. F., Hillier, I. H.: J. Chem. Soc. Faraday Trans. II *68*, 1866 (1972)
75) Hillier, I. H., Kendrick, J.: J. Chem. Soc., Faraday Trans. II *71*, 1369 (1975)
76) Vernon, G. A., Stucky, G., Carlson, T. A.: Inorg. Chem. *15*, 278 (1976)
77) Gelius, U., in: Electron spectroscopy (Shirley, D. A.) (ed.). Amsterdam: North-Holland Publ. Co. 1972, pp. 311–334
78) Pignataro, S.: Z. Naturforsch. *27a*, 816 (1972)
79) Pignataro, S., Foffani, A., Distefano, G.: Chem. Phys. Letters *20*, 350 (1973)
80) Beach, N. A., Gray, H. B.: J. Amer. Chem. Soc. *90*, 5713 (1968)
81) Basch, H.: Chem. Phys. Letters *37*, 447 (1976)
82) Mills, B. E., Shirley, D. A.: Chem. Phys. Letters *39*, 236 (1976)
83) Kim, K. S.: J. Electron Spectrosc. Rel. Phen. *3*, 217 (1974)
84) Larsson, S.: Chem. Phys. Letters *32*, 401 (1975)
85) Larsson, S.: J. Electron Spectrosc. Rel. Phen. *8*, 171 (1976)
86) Binder, H.: Z. Naturforsch. *28b*, 255 (1973)

87) Main, I. G., Marshall, J. F.: J. Phys. C: Solid State Phys. *9*, 1603 (1976)
88) Allen, G. C., Tucker, P. M.: Inorg. Chim. Acta *16*, 41 (1976)
89) Frost, D. C., McDowell, C. A., Woolsey, I. S.: Chem. Phys. Letters *17*, 320 (1972)
90) Haraguchi, H., Fujiwara, K., Fuwa, K.: Chem. Letters *1975*, 409
91) Jørgensen, C. K.: Progress Inorg. Chem. *4*, 73 (1962)
92) Orchard, A. F., Stocco, G., Thornton, G.: J. Chem. Soc., Faraday Trans. II *72*, 1045 (1976); Lazarus, M. S., Chou, T. S.: J. Chem. Phys. *64*, 3544 (1976)
93) Finn, P., Jolly, W. L.: Inorg. Chem. *11*, 893 (1972)
94) Su, C., Faller, J. W.: J. Organomet. Chem. *84*, 53 (1975)
95) Jolly, W. L., Avanzino, S. C., Schaaf, T. F., Rietz, R. R.: Unpublished data
96) Davidson, G., Riley, E. M.: Spectrochim. Acta, A *27*, 1649 (1971)
97) Werner, R. P. M., Manastyrskj, S. A.: J. Amer. Chem. Soc. *83*, 2023 (1961)
98) Durig, J. R., Marston, A. L., King, R. B., Houk, L. W.: J. Organometal. Chem. *16*, 425 (1969)
99) King, R. B., Houk, L. W.: Canadian Journal Chem. *47*, 2959 (1969)
100) Haas, H., Sheline, R. K.: J. Chem. Phys. *47*, 2996 (1967)
101) Reckziegel, A., Bigorgne, M.: J. Organometal. Chem. *3*, 341 (1965)
102) Kreiter, C. G., Fischer, E. O.: Chem. Ber. *103*, 1561 (1970)
103) Davidson, G.: Inorg. Chim. Acta *3*, 596 (1969)
104) Hieber, W., Braun, G., Beck, W.: Chem. Ber. *93*, 901 (1960)
105) Braterman, P. S., Harrill, R. W., Kaesz, H. D.: J. Amer. Chem. Soc. *89*, 2851 (1967)
106) Kaesz, H. D., Bau, R., Hendrickson, D., Smith, J. M.: J. Amer. Chem. Soc. *89*, 2844 (1967)
107) Noack, K.: J. Organometal. Chem. *12*, 181 (1968)
108) Onaka, S.: Bulletin Chem. Soc. Jap. *46*, 2444 (1973)
109) Avanzino, S. C., Jolly, W. L.: J. Am. Chem. Soc. *98*, 6505 (1976)
110) Yates, Jr., J. T., Madey, T. E., Erickson, N. E.: Surface Sci. *43*, 257 (1974)
111) Yates, Jr., J. T., Madey, T. E., Erickson, N. E., Worley, S. D.: Chem. Phys. Letters *39*, 113 (1976)
112) Iwasawa, Y., Mason, R., Textor, M., Somorjai, G. A.: Unpublished data
113) Cook, C. D., Wan, K. Y., Gelius, U., Hamrin, K., Johansson, G., Olson, E., Siegbahn, H., Nordling, C., Siegbahn, K.: J. Am. Chem. Soc. *93*, 1904 (1971)
114) Zakharova, I. A., Leites, L. A., Aleksanyan, V. T.: J. Organomet. Chem. *72*, 283 (1974)
115) Mason, R., Mingos, D. M. P., Rucci, G., Connor, J. A.: J. Chem. Soc. Dalton *1972*, 1729
116) Tolman, C. A., Riggs, W. M., Linn, W. J., King, C. M., Wendt, R. C.: Inorg. Chem. *12*, 2770 (1973)
117) Holsboer, F., Beck, W., Bartunik, H. D.: Chem. Phys. Letters *18*, 217 (1973)
118) Clarke, T. A., Gay, I. D., Mason, R.: J. Chem. Soc. Chem. Commun. *1974*, 331
119) Clarke, T. A., Gay, I. D., Law, B., Mason, R.: J. Chem. Soc. Faraday Disc. *60*, 119 (1975)
120) Mason, R., Textor, M., Iwasawa, Y., Gay, I. D.: to be published
121) Robin, M. B., Day, P.: Advances Inorg. Chem. Radiochem. *10*, 247 (1967)
122) Muetterties, E. L.: Inorg. Chem. *4*, 769 (1965)
123) Hush, N. S.: Chem. Phys. *10*, 361 (1975)
124) Cowan, D. O., Park, J., Barber, M., Swift, P.: J. Chem. Soc. Chem. Commun. *1971*, 1444
125) Burroughs, P., Hamnett, A., Orchard, A. F.: J. Chem. Soc. Dalton *1974*, 565
126) Strekas, T. C., Spiro, T. G.: Inorg. Chem. *15*, 74 (1976)
127) Beattie, J. K., Hush, N. S., Taylor, P. R.: Inorg. Chem. *15*, 992 (1976)
128) Citrin, P. H.: J. Am. Chem. Soc. *95*, 6472 (1973)
129) Cowan, D. O., LeVanda, C., Collins, R. L., Candela, G. A., Mueller-Westerhoff, U. T., Eilbracht, P.: J. Chem. Soc. Chem. Commun. *1973*, 329; LeVanda, C., Bechgaard, K., Cowan, D. O., Mueller-Westerhoff, U. T., Eilbracht, P., Candela, G. A., Collins, R. L.: J. Am. Chem. Soc. *98*, 3181 (1976)

Received October 20, 1976

Author Index Volumes 26-71

The volume numbers are printed in italics

Electrons in Fluids

The Nature of Metal-Ammonia Solutions
Editors: J. Jortner, N. R. Kestner

With contributions by: J. V. Acrivos, C. L. van Antwerp, K. Bar-Eli, J. Bellon[illegible], H. Boll, D. E. Bowen, K. G. Breitschwerdt, R. G. Brown, M. H. Cohen, R. M. Cotts, P. Damay, T. David, H. T. Davis, M. G. DeBacker, B. DeBettignies, P. Delahay, R. R. Dewald, L. M. Dorfman, J. L. Dye, U. Even, A. Gaathon, A. N. Garroway, T. H. Geballe, A. Gedanken, W. S. Glaunsinger, F. Hensel, J. A. Hamilton, K. Ichikawa, R. L. Jones, J. Jortner, F. Y. Jou, N. R. Kestner, W. H. Koehler, J. J. Lagowski, C. Lambert, J. P. Lelieur, G. Lepoutre, J. Logan, S. F. Meyer, S. Nehari, H. Radscheit, B. Raz, P. M. Rentzepis, J. H. Roberts, P. F. Rusch, E. Saito, P. D. Schettler, U. Schindewolf, L. D. Schmidt, M. J. Sienko, B. L. Smith, J. D. Spear, J. C. Thompson, J. E. Thilly, I. Webman, S. Zolotov.

271 figures (including two color plates), 59 tables. XII, 439 pages. 1973
ISBN 3–540–06310–2

This full and up-to-date account reveals the interrelationships between the chemistry and physics of electrons in disordered media in a way that appeals to physicists as well as chemists. The interdisciplinary nature of this field and related areas is clear from the range of contributors, which includes both theoretical and experimental chemists and physicists.

From the Content: Theory of Electrons in Polar Fluids. –
Metal-Ammonia Solutions: The Dilute Region. –
Metal Solutions in Amines and Ethers. –
Ultrafast Optical Processes. –
Metal-Ammonia Solutions: Transition Range. –
The Electronic Structures of Disordered Materials. –
Concentrated M–NH_3 Solutions. –
Strange Magnetic Behavior and Phase Relations of Metal-Ammonia Compounds. –
Metallic Vapors. –
Mobility Studies of Excess Electrons in Nonpolar Hydrocarbons. –
Optical Absorption Spectrum of the Solvated Electron in Ethers and in Binary Liquid Systems.

Springer-Verlag
Berlin Heidelberg New York

NMR
Basic Principles and Progress
Grundlagen und Fortschritte

Editors: P. Diehl, E. Fluck, R. Kosfeld

Volume 10
Van der Waals Forces and Shielding Effects
13 figures, 46 tables.
II, 118 pages. 1975
ISBN 3–540–07340–X

Volume 11
M. Mehring
High Resolution NMR Spectroscopy in Solids
104 figures. XI, 246 pages. 1976
ISBN 3–540–07704–9

Volume 12
B. Lindman, S. Forsén
Chlorine, Bromine and Iodine NMR
Physico-Chemical and Biological Applications
74 figures, 29 tables.
XIV, 368 pages. 1976
ISBN 3–540–07725–1

Volume 13
Introductory Essays
Editor: M. M. Pintar
48 figures. XI, 154 pages. 1976
ISBN 3–540–07754–5

Structure and Bonding

Editors: J. D. Dunitz, P. Hemmerich, J. A. Ibers, C. K. Jørgensen, J. B. Neilands, D. Reinen, R. J. P. Williams

Volume 29
Biochemistry
51 figures, 48 tables.
IV, 219 pages. 1976
ISBN 3–540–07886–X

Volume 30
Rare Earths
104 figures, 18 tables.
IV, 197 pages. 1976
ISBN 3–540–07887–8

Volume 31
Bonding and Compounds of Less Abundant Metals
40 figures, 6 tables.
IV, 112 pages. 1976
ISBN 3–540–07964–5

Volume 32
Novel Chemical Effects of Electronic Behaviour
31 figures, 16 tables.
IV, 171 pages. 1977
ISBN 3–540–08014–7

Springer-Verlag
Berlin Heidelberg New York